PRAISE FOR THE *MANGA GUIDE* SERIES

"Highly recommended."
—CHOICE MAGAZINE ON *THE MANGA GUIDE TO DATABASES*

"Stimulus for the next generation of scientists."
—SCIENTIFIC COMPUTING ON *THE MANGA GUIDE TO MOLECULAR BIOLOGY*

"A great fit of form and subject. Recommended."
—OTAKU USA MAGAZINE ON *THE MANGA GUIDE TO PHYSICS*

"The art is charming and the humor engaging. A fun and fairly painless lesson on what many consider to be a less-than-thrilling subject."
—SCHOOL LIBRARY JOURNAL ON *THE MANGA GUIDE TO STATISTICS*

"This is really what a good math text should be like. Unlike the majority of books on subjects like statistics, it doesn't just present the material as a dry series of pointless-seeming formulas. It presents statistics as something *fun*, and something enlightening."
—GOOD MATH, BAD MATH ON *THE MANGA GUIDE TO STATISTICS*

"I found the cartoon approach of this book so compelling and its story so endearing that I recommend that every teacher of introductory physics, in both high school and college, consider using it."
—AMERICAN JOURNAL OF PHYSICS ON *THE MANGA GUIDE TO PHYSICS*

"A single tortured cry will escape the lips of every thirty-something biochem major who sees *The Manga Guide to Molecular Biology*: 'Why, oh why couldn't this have been written when I was in college?'"
—THE SAN FRANCISCO EXAMINER

"A lot of fun to read. The interactions between the characters are lighthearted, and the whole setting has a sort of quirkiness about it that makes you keep reading just for the joy of it."
—HACK A DAY ON *THE MANGA GUIDE TO ELECTRICITY*

"*The Manga Guide to Databases* was the most enjoyable tech book I've ever read."
—RIKKI KITE, LINUX PRO MAGAZINE

"For parents trying to give their kids an edge or just for kids with a curiosity about their electronics, *The Manga Guide to Electricity* should definitely be on their bookshelves."
—SACRAMENTO BOOK REVIEW

WOW!

"This is a solid book and I wish there were more like it in the IT world."
—SLASHDOT ON *THE MANGA GUIDE TO DATABASES*

"*The Manga Guide to Electricity* makes accessible a very intimidating subject, letting the reader have fun while still delivering the goods."
—GEEKDAD BLOG, WIRED.COM

"If you want to introduce a subject that kids wouldn't normally be very interested in, give it an amusing storyline and wrap it in cartoons."
—MAKE ON *THE MANGA GUIDE TO STATISTICS*

"This book does exactly what it is supposed to: offer a fun, interesting way to learn calculus concepts that would otherwise be extremely bland to memorize."
—DAILY TECH ON *THE MANGA GUIDE TO CALCULUS*

"The art is fantastic, and the teaching method is both fun and educational."
—ACTIVE ANIME ON *THE MANGA GUIDE TO PHYSICS*

"An awfully fun, highly educational read."
—FRAZZLEDDAD ON *THE MANGA GUIDE TO PHYSICS*

"Makes it possible for a 10-year-old to develop a decent working knowledge of a subject that sends most college students running for the hills."
—SKEPTICBLOG ON *THE MANGA GUIDE TO MOLECULAR BIOLOGY*

"This book is by far the best book I have read on the subject. I think this book absolutely rocks and recommend it to anyone working with or just interested in databases."
—GEEK AT LARGE ON *THE MANGA GUIDE TO DATABASES*

"The book purposefully departs from a traditional physics textbook and it does it very well."
—DR. MARINA MILNER-BOLOTIN, RYERSON UNIVERSITY ON *THE MANGA GUIDE TO PHYSICS*

"Kids would be, I think, much more likely to actually pick this up and find out if they are interested in statistics as opposed to a regular textbook."
—GEEK BOOK ON *THE MANGA GUIDE TO STATISTICS*

"*The Manga Guide to Statistics* offers a visualization of statistics that can't be found in any mere textbook."
—ANIME 3000

"A great introduction for readers of any age, and an exemplar of technical communication."
—LINUX USERS OF VICTORIA ON *THE MANGA GUIDE TO ELECTRICITY*

THE MANGA GUIDE™ TO RELATIVITY

THE MANGA GUIDE™ TO
RELATIVITY

HIDEO NITTA
MASAFUMI YAMAMOTO
KEITA TAKATSU
TREND-PRO CO., LTD.

The Manga Guide to Relativity is a translation of the Japanese original, Manga de wakaru soutaiseiriron, published by Ohmsha, Ltd. of Tokyo, Japan. © 2009 by Hideo Nitta, Masafumi Yamamoto, and TREND-PRO Co., Ltd.

This English edition is co-published by No Starch Press, Inc. and Ohmsha, Ltd.

15 14 13 12 11   1 2 3 4 5 6 7 8 9

ISBN-10: 1-59327-272-3
ISBN-13: 978-1-59327-272-2

Publisher: William Pollock
Supervising Editor: Hideo Nitta
Author: Masafumi Yamamoto
Illustrator: Keita Takatsu
Producer: TREND-PRO Co., Ltd.
Production Editor: Serena Yang
Developmental Editor: Tyler Ortman
Translator: Arnie Rusoff
Technical Reviewers: David Issadore and John Roeder
Compositor: Riley Hoffman
Copyeditor: Paula L. Fleming
Proofreader: Serena Yang
Indexer: Sarah Schott

For information on book distributors or translations, please contact No Starch Press, Inc. directly:
No Starch Press, Inc.
38 Ringold Street, San Francisco, CA 94103
phone: 415.863.9900; fax: 415.863.9950; info@nostarch.com; http://www.nostarch.com/

Library of Congress Cataloging-in-Publication Data

Nitta, Hideo, 1957-
 [Manga de wakaru soutaiseiriron. English]
 The manga guide to relativity / Hideo Nitta, Masafumi Yamamoto, Keita Takatsu ; Trend-pro Co., Ltd. -- English ed.
     p. cm.
 Includes index.
 ISBN-13: 978-1-59327-272-2
 ISBN-10: 1-59327-272-3
 1. Relativity (Physics)--Comic books, strips, etc. 2. Graphic novels. I. Yamamoto, Masafumi, 1947- II. Takatsu, Keita. III. Trend-pro Co. IV. Title.
 QC173.57.N5813 2010
 530.11--dc22

                                    2010038111

TABLE OF CONTENTS

4
WHAT IS GENERAL RELATIVITY

PREFACE

Welcome to the world of relativity!

Everyone wonders what relativity is all about. Because the theory of relativity predicts phenomena that seem unbelievable in our everyday lives (such as the slowing of time and the contraction of the length of an object), it can seem like mysterious magic.

Despite its surprising, counterintuitive predictions, Einstein's theory of relativity has been confirmed many times over with countless experiments by modern physicists. Relativity and the equally unintuitive quantum mechanics are indispensable tools for understanding the physical world.

In Newton's time, when physicists considered velocities much smaller than the speed of light, it was not a problem to think that the measurement of motion, that is, space and time, were independent, permanent, and indestructible absolutes. However, by the end of the 19th century, precise measurements of the speed of light combined with developments in the study of electromagnetism had set the stage for the discovery of relativity. As a result, time and space, which had always been considered to be independent and absolute, had to be reconsidered.

That's when Einstein arrived on the scene. Einstein proposed that time and space were in fact relative. He discarded the idea that space and time were absolute and considered that they vary together, so that the speed of light is always constant.

This radical insight created a controversy just as Galileo's claim that Earth orbited the Sun (and not vice versa) shocked his peers. However, once we ventured into space, it was obvious that Earth was indeed moving.

In a similar way, relativity has given us a more accurate understanding of concepts regarding the space-time in which we are living. In other words, relativity is the result of asking what is *actually* happening in our world rather than saying our world *should be* a particular way.

Although this preface may seem a little difficult, I hope you will enjoy the mysteries of relativity in a manga world together with Minagi and his teacher, Miss Uraga. Finally, I'd like to express my deep gratitude to everyone in the development bureau at Ohmsha; re_akino, who toiled over the scenario; and Mr. Keita Takatsu, who converted it into such an interesting manga.

Well, then. Let's jump into the world of relativity.

MASAFUMI YAMAMOTO
JUNE 2009

PROLOGUE

OUTRAGEOUS CLOSING CEREMONY

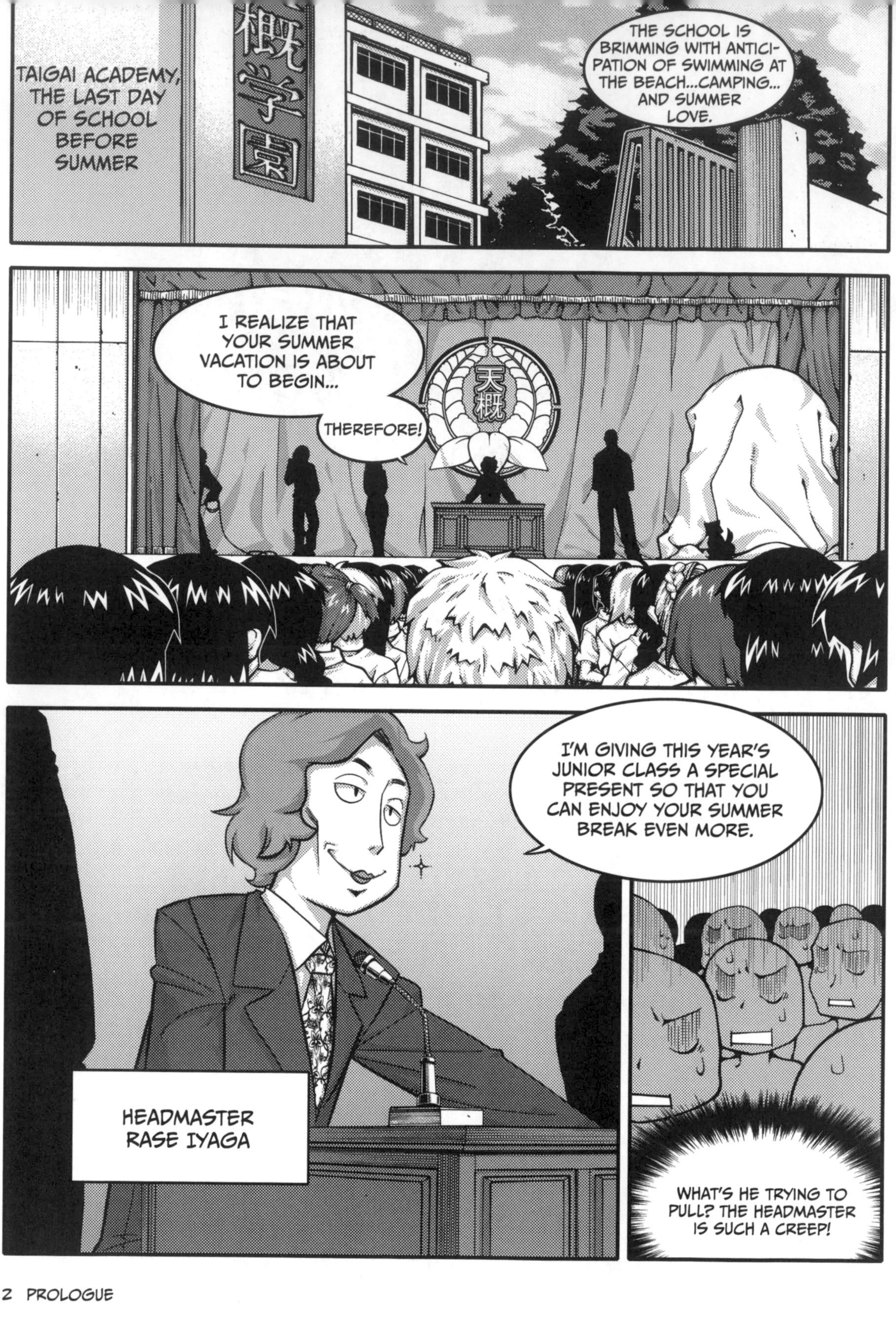
TAIGAI ACADEMY, THE LAST DAY OF SCHOOL BEFORE SUMMER
概学園
THE SCHOOL IS BRIMMING WITH ANTICI-PATION OF SWIMMING AT THE BEACH...CAMPING... AND SUMMER LOVE.
I REALIZE THAT YOUR SUMMER VACATION IS ABOUT TO BEGIN...
THEREFORE!
天概
I'M GIVING THIS YEAR'S JUNIOR CLASS A SPECIAL PRESENT SO THAT YOU CAN ENJOY YOUR SUMMER BREAK EVEN MORE.
HEADMASTER RASE IYAGA
WHAT'S HE TRYING TO PULL? THE HEADMASTER IS SUCH A CREEP!

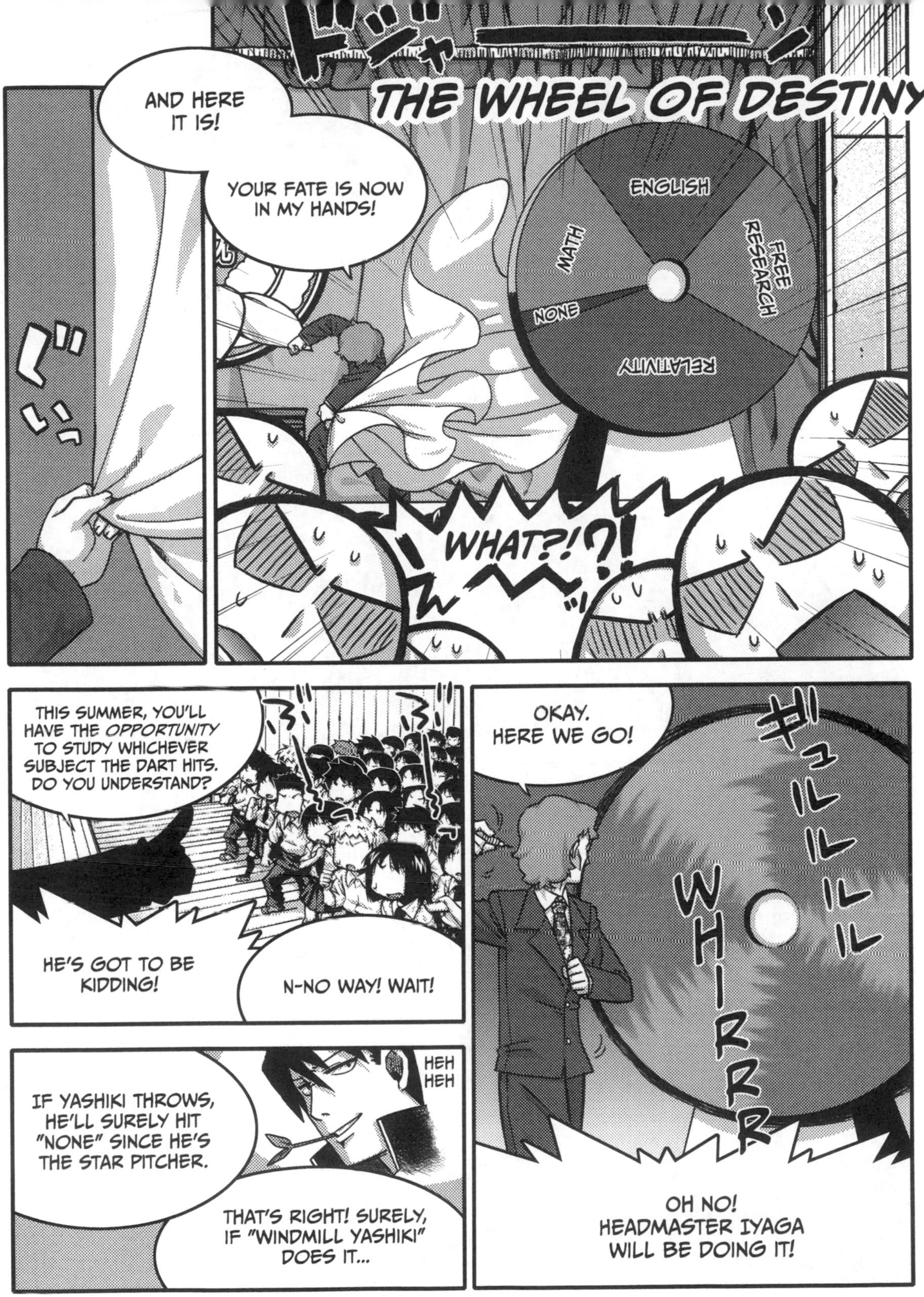
THE WHEEL OF DESTINY!
AND HERE IT IS!
YOUR FATE IS NOW IN MY HANDS!
ENGLISH
FREE RESEARCH
RELATIVITY
NONE
MATH
WHAT?!?!
THIS SUMMER, YOU'LL HAVE THE *OPPORTUNITY* TO STUDY WHICHEVER SUBJECT THE DART HITS. DO YOU UNDERSTAND?
HE'S GOT TO BE KIDDING!
N-NO WAY! WAIT!
IF YASHIKI THROWS, HE'LL SURELY HIT "NONE" SINCE HE'S THE STAR PITCHER.
HEH HEH
THAT'S RIGHT! SURELY, IF "WINDMILL YASHIKI" DOES IT...
OKAY. HERE WE GO!
WHIRRR
OH NO! HEADMASTER IYAGA WILL BE DOING IT!

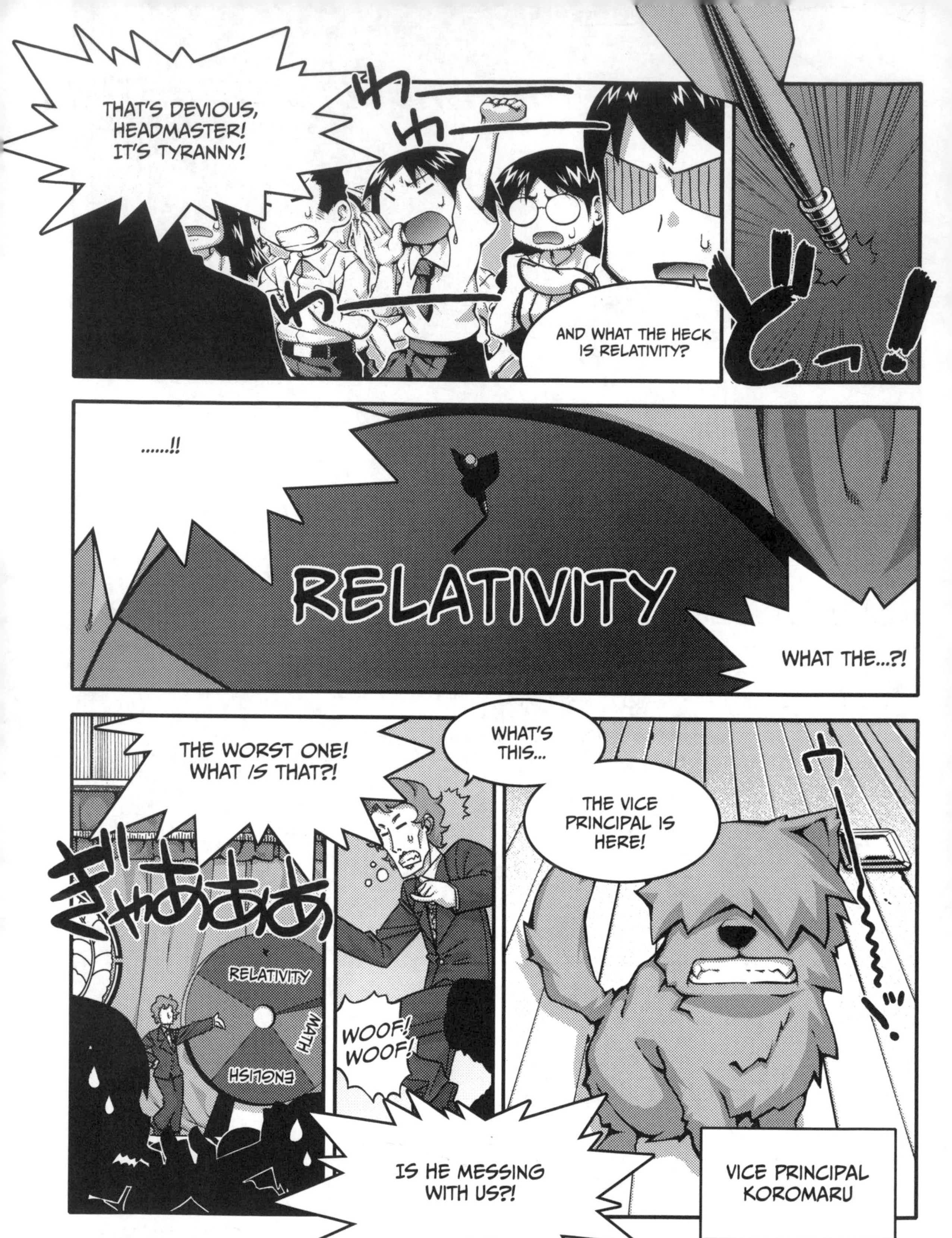
THAT'S DEVIOUS, HEADMASTER! IT'S TYRANNY!
わー
わー
わ
AND WHAT THE HECK IS RELATIVITY?
どーっ!
......!!
RELATIVITY
WHAT THE...?!
THE WORST ONE! WHAT IS THAT?!
ぎゃあああ
RELATIVITY
MATH
ENGLISH
WHAT'S THIS...
WOOF! WOOF!
THE VICE PRINCIPAL IS HERE!
ウー...!
IS HE MESSING WITH US?!
VICE PRINCIPAL KOROMARU

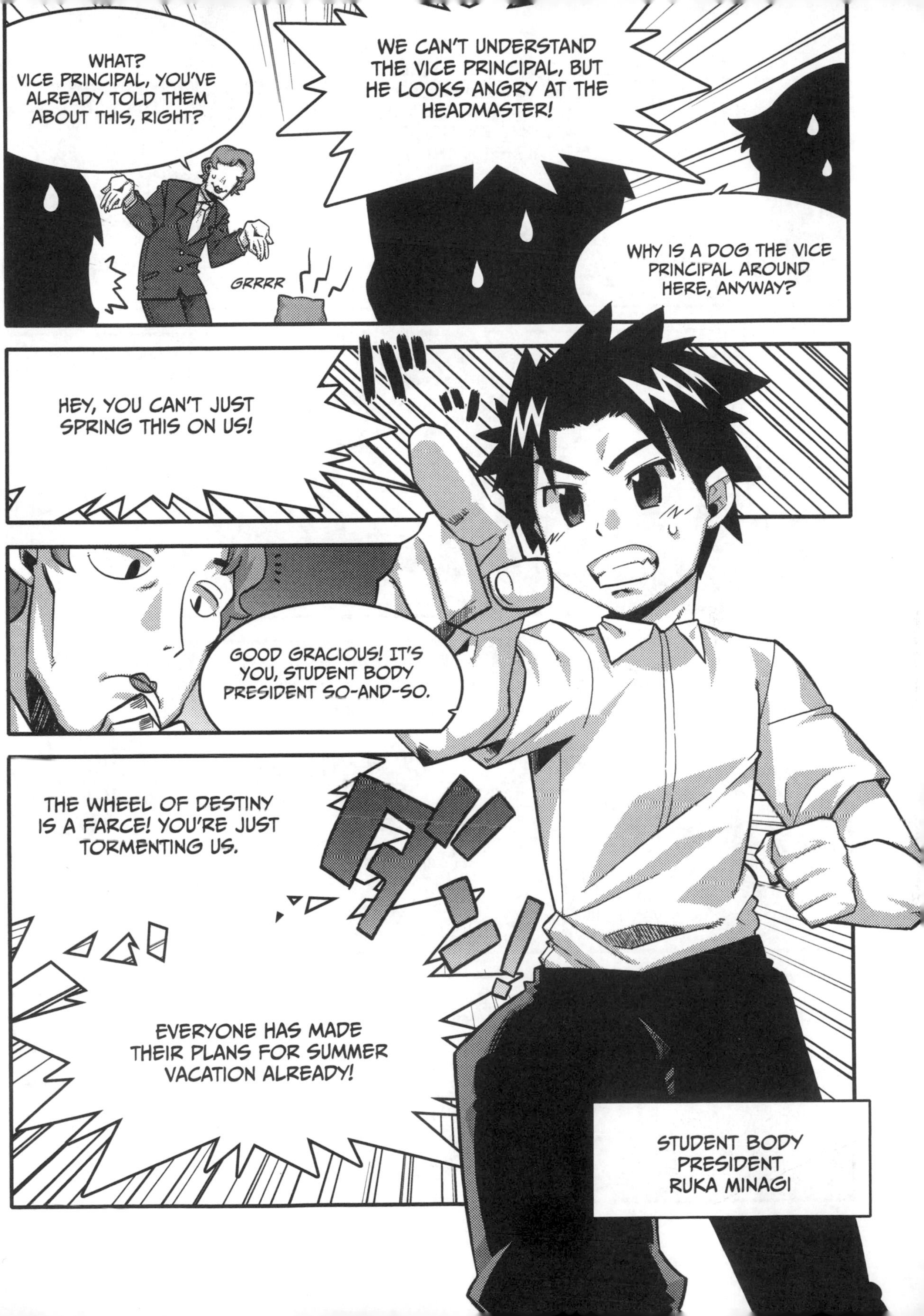
WHAT? VICE PRINCIPAL, YOU'VE ALREADY TOLD THEM ABOUT THIS, RIGHT?
GRRRR
WE CAN'T UNDERSTAND THE VICE PRINCIPAL, BUT HE LOOKS ANGRY AT THE HEADMASTER!
WHY IS A DOG THE VICE PRINCIPAL AROUND HERE, ANYWAY?
HEY, YOU CAN'T JUST SPRING THIS ON US!
GOOD GRACIOUS! IT'S YOU, STUDENT BODY PRESIDENT SO-AND-SO.
THE WHEEL OF DESTINY IS A FARCE! YOU'RE JUST TORMENTING US.
EVERYONE HAS MADE THEIR PLANS FOR SUMMER VACATION ALREADY!
STUDENT BODY PRESIDENT RUKA MINAGI

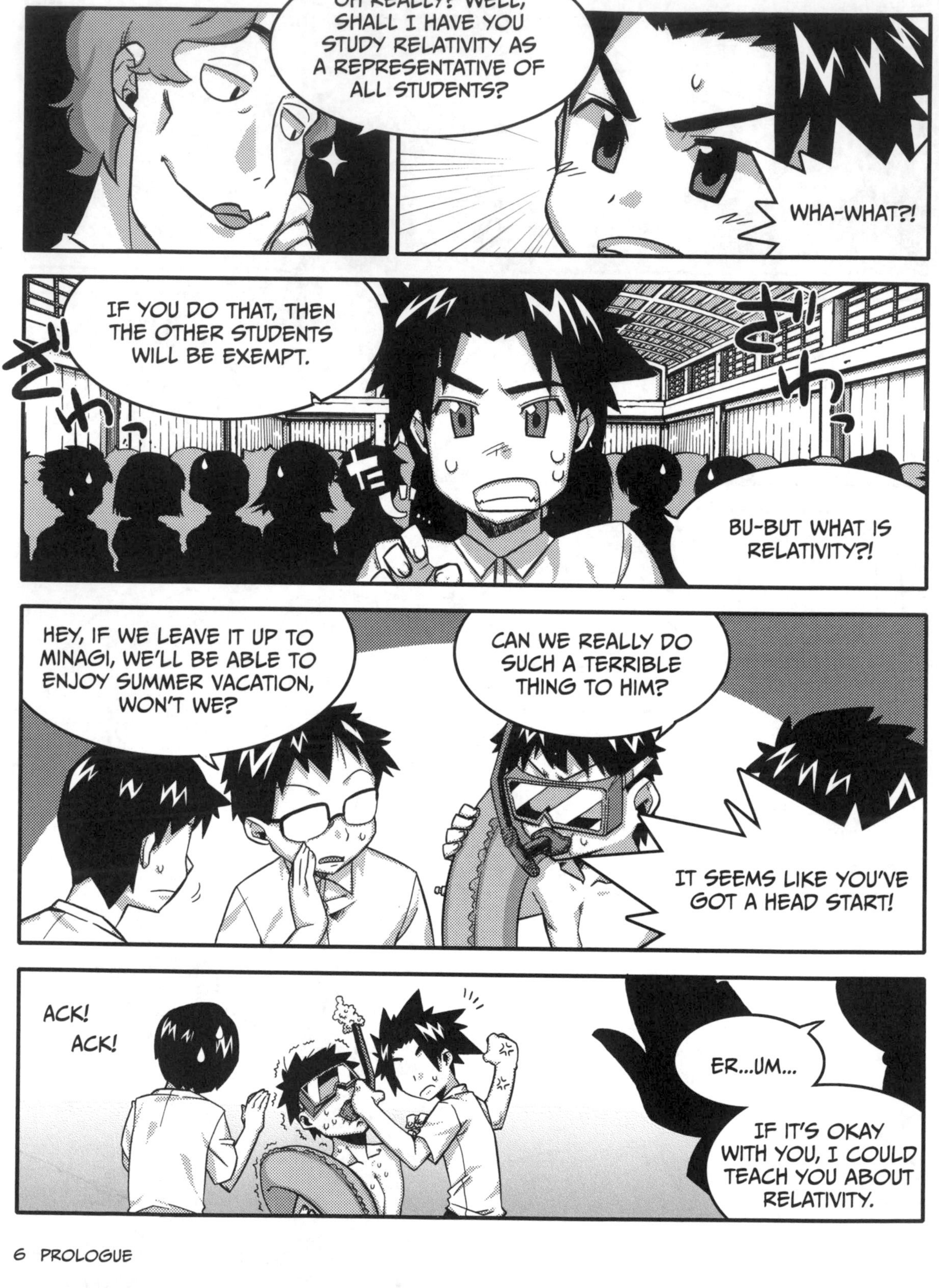
OH REALLY? WELL, SHALL I HAVE YOU STUDY RELATIVITY AS A REPRESENTATIVE OF ALL STUDENTS?
WHA-WHAT?!
ざわっ
IF YOU DO THAT, THEN THE OTHER STUDENTS WILL BE EXEMPT.
ざわっ
BU-BUT WHAT IS RELATIVITY?!
HEY, IF WE LEAVE IT UP TO MINAGI, WE'LL BE ABLE TO ENJOY SUMMER VACATION, WON'T WE?
CAN WE REALLY DO SUCH A TERRIBLE THING TO HIM?
IT SEEMS LIKE YOU'VE GOT A HEAD START!
ACK!
ACK!
ER...UM...
IF IT'S OKAY WITH YOU, I COULD TEACH YOU ABOUT RELATIVITY.

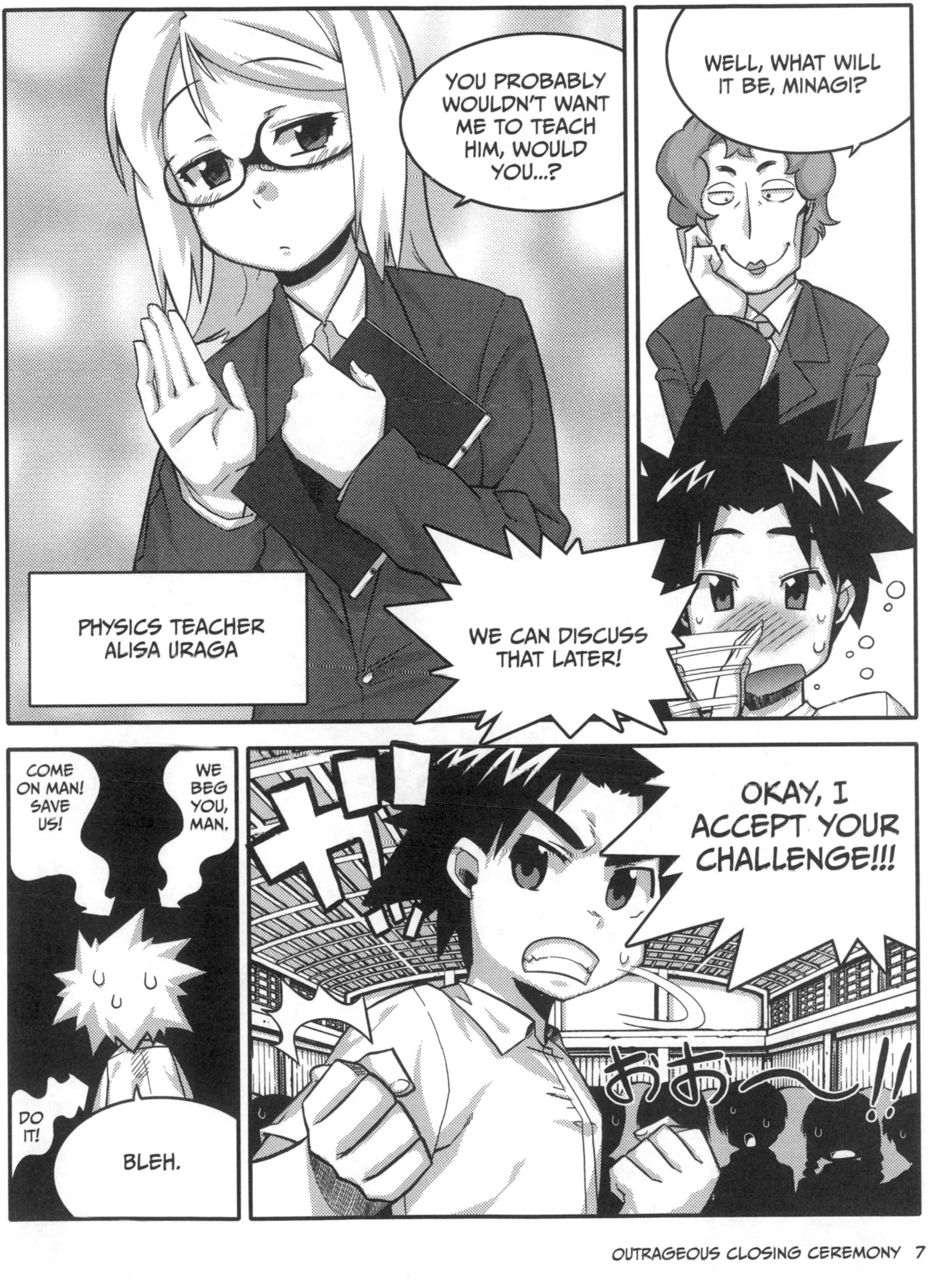
YOU PROBABLY WOULDN'T WANT ME TO TEACH HIM, WOULD YOU...?
PHYSICS TEACHER ALISA URAGA
WELL, WHAT WILL IT BE, MINAGI?
WE CAN DISCUSS THAT LATER!
COME ON MAN! SAVE US!
WE BEG YOU, MAN.
DO IT!
BLEH.
OKAY, I ACCEPT YOUR CHALLENGE!!!
おおーー!!

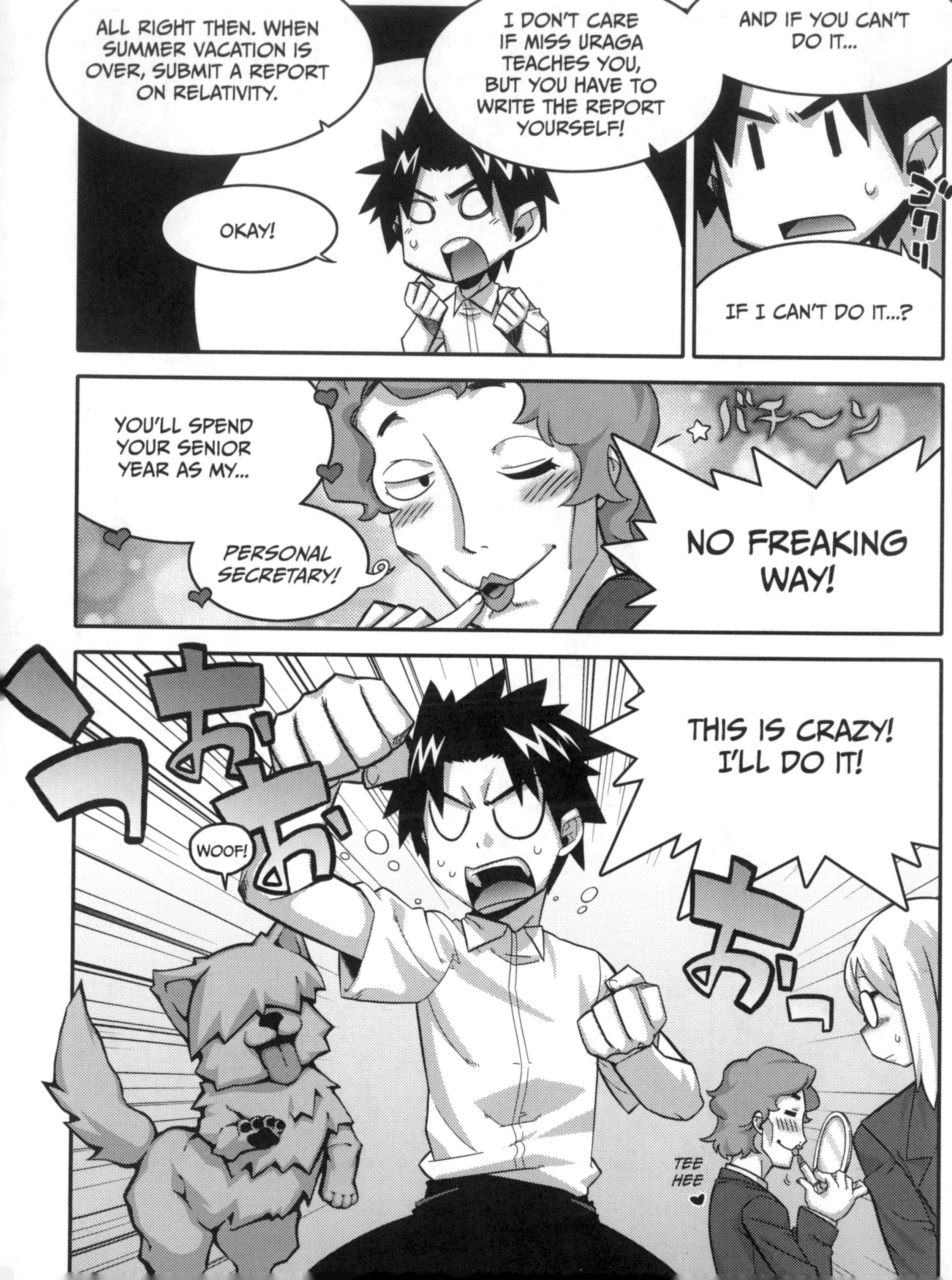
ALL RIGHT THEN. WHEN SUMMER VACATION IS OVER, SUBMIT A REPORT ON RELATIVITY.
OKAY!
I DON'T CARE IF MISS URAGA TEACHES YOU, BUT YOU HAVE TO WRITE THE REPORT YOURSELF!
AND IF YOU CAN'T DO IT...
IF I CAN'T DO IT...?
YOU'LL SPEND YOUR SENIOR YEAR AS MY...
PERSONAL SECRETARY!
NO FREAKING WAY!
WOOF!
THIS IS CRAZY! I'LL DO IT!
TEE HEE

WHAT IS RELATIVITY?

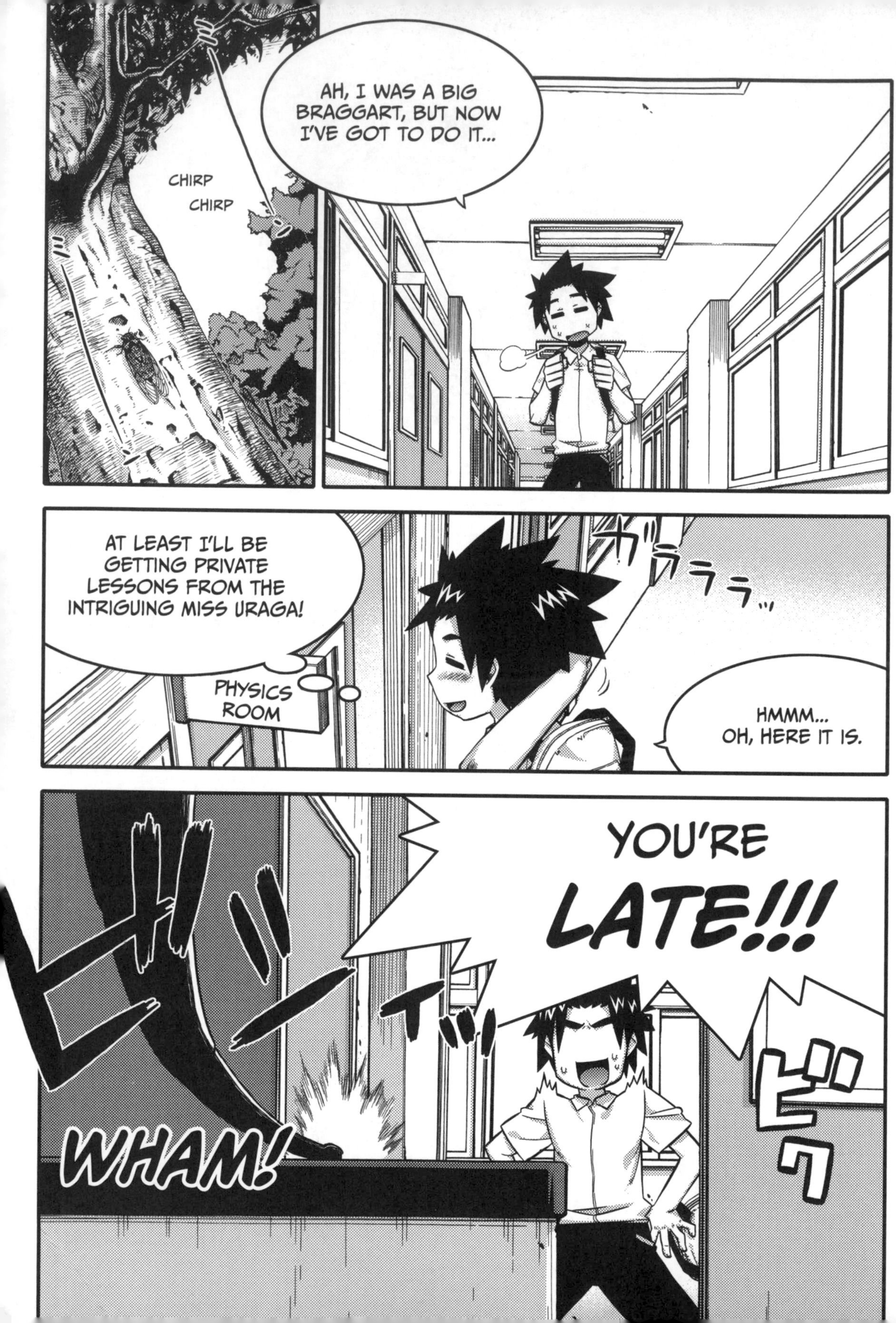
CHIRP
CHIRP
AH, I WAS A BIG BRAGGART, BUT NOW I'VE GOT TO DO IT...
AT LEAST I'LL BE GETTING PRIVATE LESSONS FROM THE INTRIGUING MISS URAGA!
PHYSICS ROOM
HMMM... OH, HERE IT IS.
WHAM!
YOU'RE LATE!!!

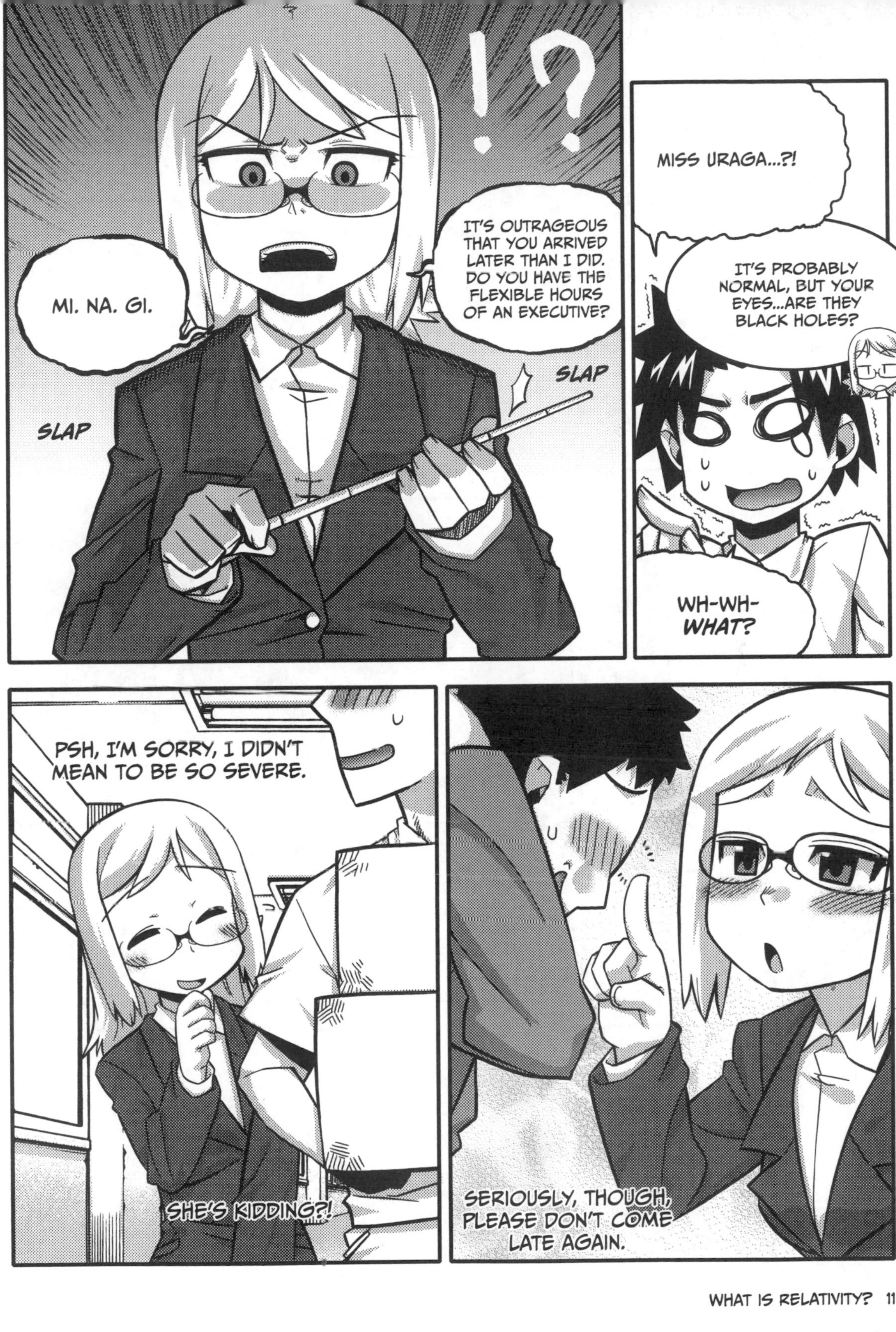
!?
MI. NA. GI.
IT'S OUTRAGEOUS THAT YOU ARRIVED LATER THAN I DID. DO YOU HAVE THE FLEXIBLE HOURS OF AN EXECUTIVE?
SLAP
SLAP
MISS URAGA...?!
IT'S PROBABLY NORMAL, BUT YOUR EYES...ARE THEY BLACK HOLES?
WH-WH-WHAT?
PSH, I'M SORRY, I DIDN'T MEAN TO BE SO SEVERE.
SHE'S KIDDING?!
SERIOUSLY, THOUGH, PLEASE DON'T COME LATE AGAIN.

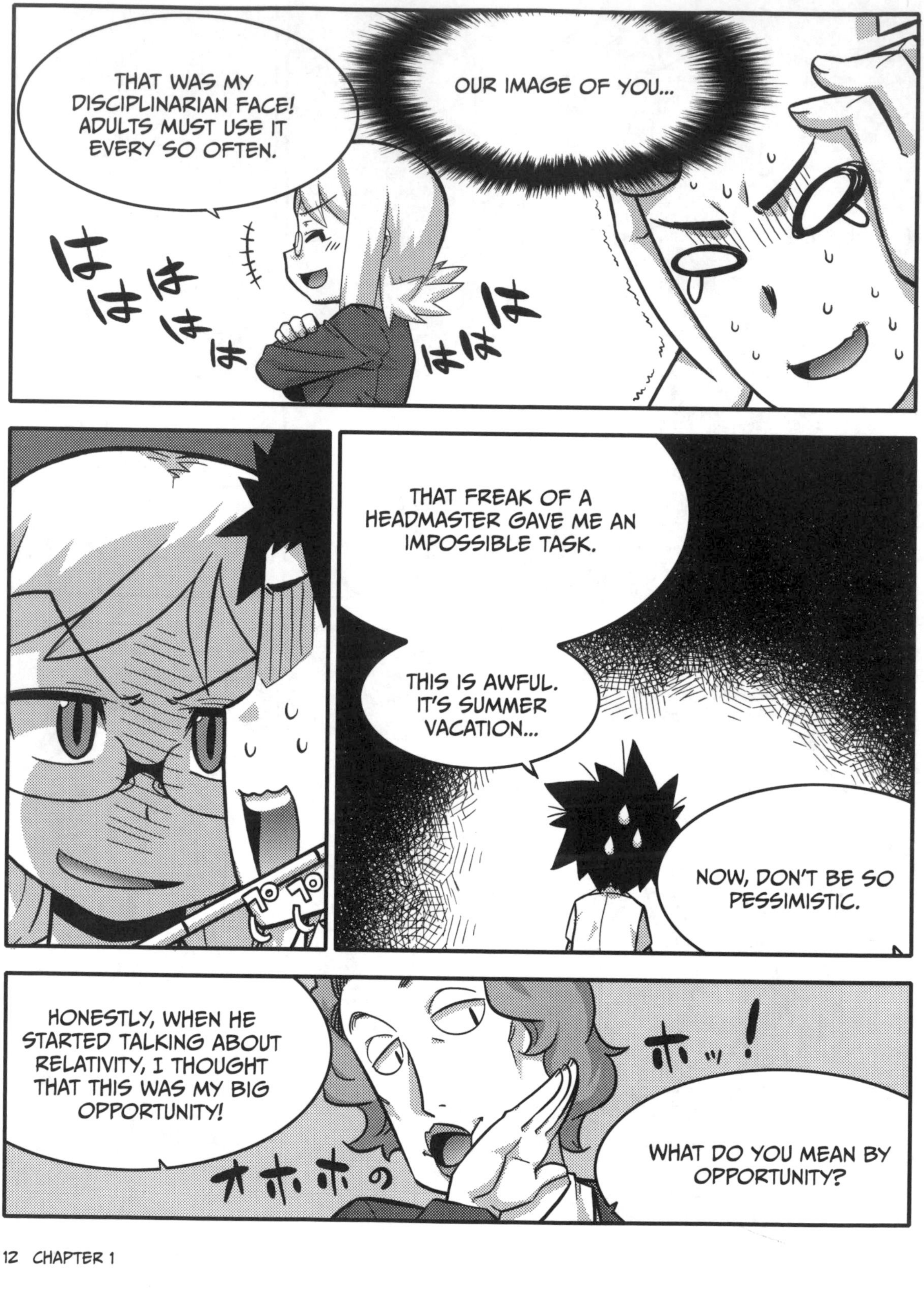

THAT WAS MY DISCIPLINARIAN FACE! ADULTS MUST USE IT EVERY SO OFTEN.
ははははは
OUR IMAGE OF YOU...
ははは
THAT FREAK OF A HEADMASTER GAVE ME AN IMPOSSIBLE TASK.
THIS IS AWFUL. IT'S SUMMER VACATION...
NOW, DON'T BE SO PESSIMISTIC.
HONESTLY, WHEN HE STARTED TALKING ABOUT RELATIVITY, I THOUGHT THAT THIS WAS MY BIG OPPORTUNITY!
ホッ!
オホホの
WHAT DO YOU MEAN BY OPPORTUNITY?

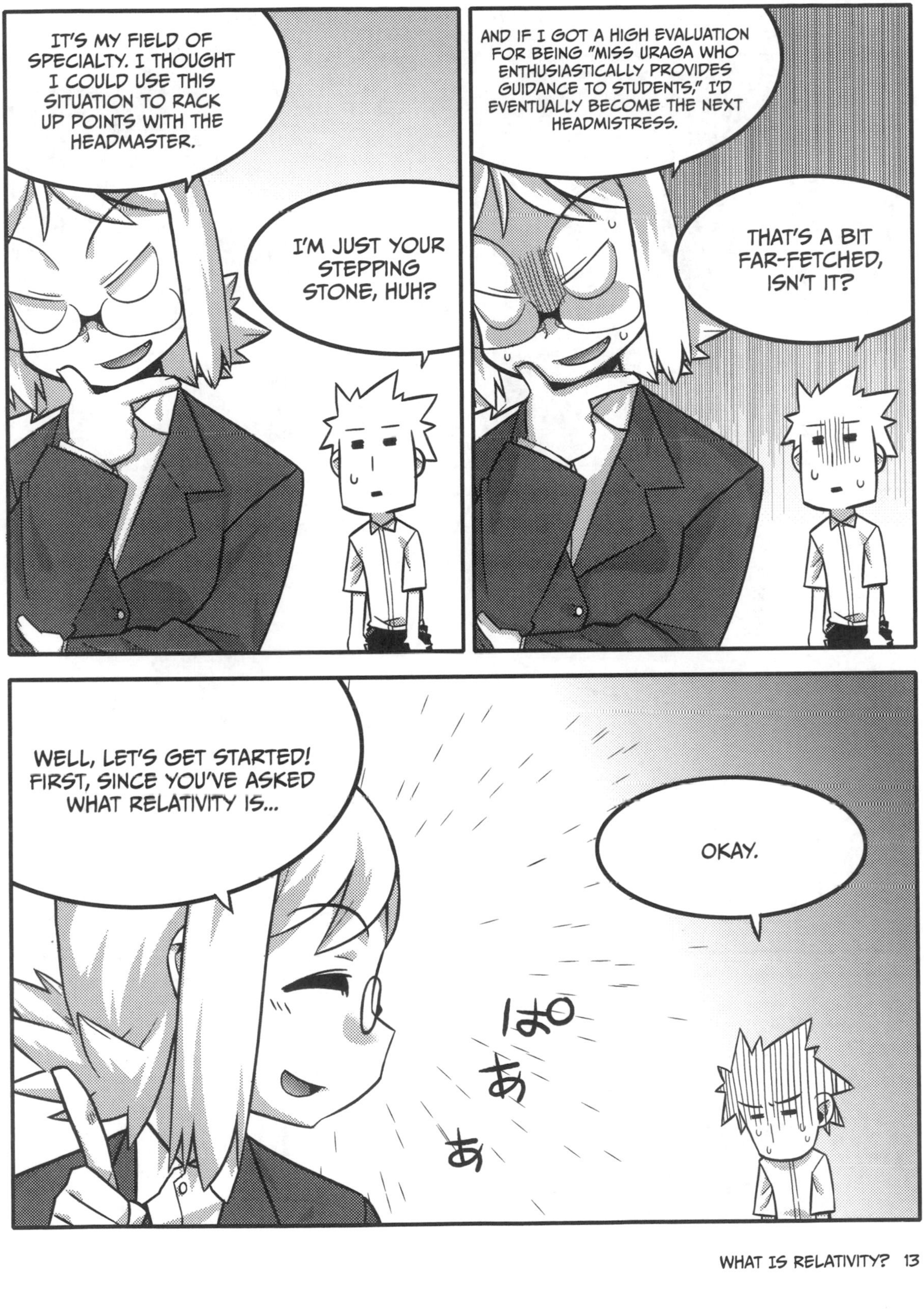
IT'S MY FIELD OF SPECIALTY. I THOUGHT I COULD USE THIS SITUATION TO RACK UP POINTS WITH THE HEADMASTER.
I'M JUST YOUR STEPPING STONE, HUH?
AND IF I GOT A HIGH EVALUATION FOR BEING "MISS URAGA WHO ENTHUSIASTICALLY PROVIDES GUIDANCE TO STUDENTS," I'D EVENTUALLY BECOME THE NEXT HEADMISTRESS.
THAT'S A BIT FAR-FETCHED, ISN'T IT?
WELL, LET'S GET STARTED! FIRST, SINCE YOU'VE ASKED WHAT RELATIVITY IS...
OKAY.
ぱあぁ

1. WHAT IS RELATIVITY?

SPECIAL   GENERAL

THERE ARE TWO TYPES OF RELATIVITY. ONE IS *SPECIAL RELATIVITY*, AND THE OTHER IS *GENERAL RELATIVITY*.

GENERAL RELATIVITY EXTENDED SPECIAL RELATIVITY.

SHOULDN'T A GENERAL THEORY COME BEFORE A SPECIALIZED ONE?

SPECIAL RELATIVITY IS "SPECIAL" BECAUSE IT IS A SIMPLIFICATION OF GENERAL RELATIVITY AND IS ONLY TRUE IN THE SPECIAL CASES WHEN THE EFFECTS OF GRAVITY AND ACCELERATION CAN BE SAFELY IGNORED.

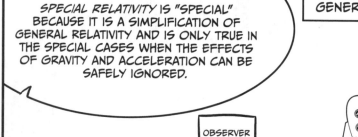

OBSERVER

SPECIAL RELATIVITY

GENERAL RELATIVITY

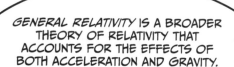

OBSERVER

GENERAL RELATIVITY IS A BROADER THEORY OF RELATIVITY THAT ACCOUNTS FOR THE EFFECTS OF BOTH ACCELERATION AND GRAVITY.

IS IT SIMPLER WHEN GRAVITY OR ACCELERATION IS NOT CONSIDERED?

WELL, SURE IT IS.

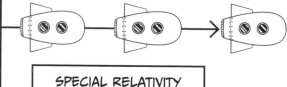

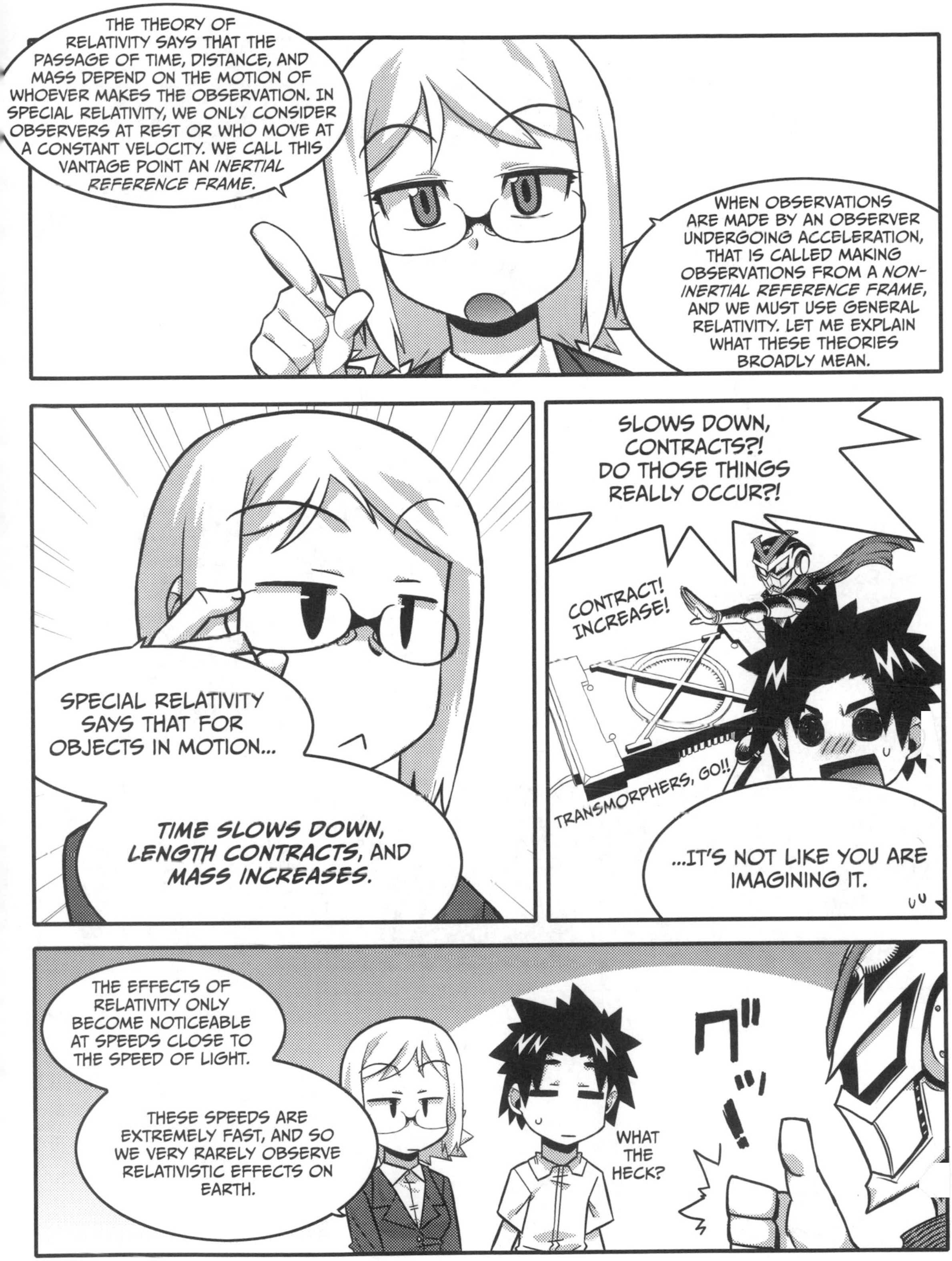
THE THEORY OF RELATIVITY SAYS THAT THE PASSAGE OF TIME, DISTANCE, AND MASS DEPEND ON THE MOTION OF WHOEVER MAKES THE OBSERVATION. IN SPECIAL RELATIVITY, WE ONLY CONSIDER OBSERVERS AT REST OR WHO MOVE AT A CONSTANT VELOCITY. WE CALL THIS VANTAGE POINT AN *INERTIAL REFERENCE FRAME.*
WHEN OBSERVATIONS ARE MADE BY AN OBSERVER UNDERGOING ACCELERATION, THAT IS CALLED MAKING OBSERVATIONS FROM A *NON-INERTIAL REFERENCE FRAME,* AND WE MUST USE GENERAL RELATIVITY. LET ME EXPLAIN WHAT THESE THEORIES BROADLY MEAN.
SPECIAL RELATIVITY SAYS THAT FOR OBJECTS IN MOTION...
TIME SLOWS DOWN, LENGTH CONTRACTS, AND ***MASS INCREASES.***
SLOWS DOWN, CONTRACTS?! DO THOSE THINGS REALLY OCCUR?!
CONTRACT! INCREASE!
TRANSMORPHERS, GO!!
...IT'S NOT LIKE YOU ARE IMAGINING IT.
THE EFFECTS OF RELATIVITY ONLY BECOME NOTICEABLE AT SPEEDS CLOSE TO THE SPEED OF LIGHT.
THESE SPEEDS ARE EXTREMELY FAST, AND SO WE VERY RARELY OBSERVE RELATIVISTIC EFFECTS ON EARTH.
WHAT THE HECK?

GENERAL RELATIVITY SAYS THAT AN OBJECT WITH MASS *CREATES* GRAVITY BY AFFECTING TIME AND SPACE.

WHAT IS GRAVITY?

RELATIVITY FLASH!

I SEE. FOR EXAMPLE, LIGHT...

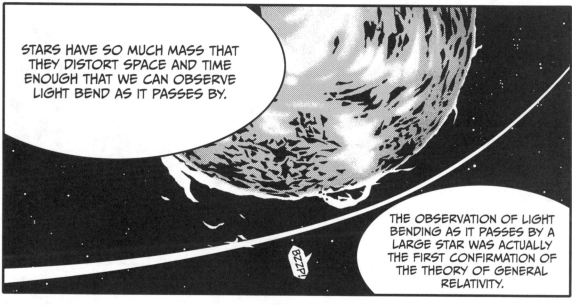

STARS HAVE SO MUCH MASS THAT THEY DISTORT SPACE AND TIME ENOUGH THAT WE CAN OBSERVE LIGHT BEND AS IT PASSES BY.

BZZP!

THE OBSERVATION OF LIGHT BENDING AS IT PASSES BY A LARGE STAR WAS ACTUALLY THE FIRST CONFIRMATION OF THE THEORY OF GENERAL RELATIVITY.

NOW, SINCE GENERAL RELATIVITY IS MORE ADVANCED AND DIFFICULT...

LET'S PROCEED BY FOCUSING OUR DISCUSSION ON SPECIAL RELATIVITY. YOU SHOULD APPRECIATE THIS.

OKAAAAY...

BECAUSE IF I DON'T UNDERSTAND, I'LL REALLY BE IN HOT WATER....

2. GALILEAN PRINCIPLE OF RELATIVITY AND NEWTONIAN MECHANICS

LET'S BEGIN WITH SOME HISTORICAL BACKGROUND SO YOU CAN UNDERSTAND RELATIVITY A LITTLE BETTER.

HISTORICAL BACKGROUND?

IT WILL PROBABLY BE EASIER FOR YOU TO UNDERSTAND IF I TELL YOU HOW THE THEORY OF RELATIVITY ORIGINATED.

BY THE WAY, SINCE I'LL GLOSS OVER SOME DETAILS, I HOPE YOU'LL FORGIVE ME IF MY EXPLANATION LACKS A LITTLE SCIENTIFIC PRECISION.

WELL, I GUESS IT'S OKAY AS LONG AS I GET THE GENERAL IDEA.

FIRST, MORE THAN 300 YEARS BEFORE EINSTEIN MADE HIS APPEARANCE...

Galileo Galilei

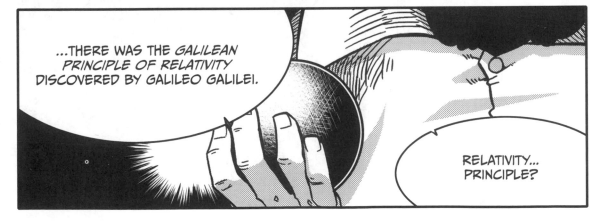

...THERE WAS THE *GALILEAN PRINCIPLE OF RELATIVITY* DISCOVERED BY GALILEO GALILEI.

RELATIVITY... PRINCIPLE?

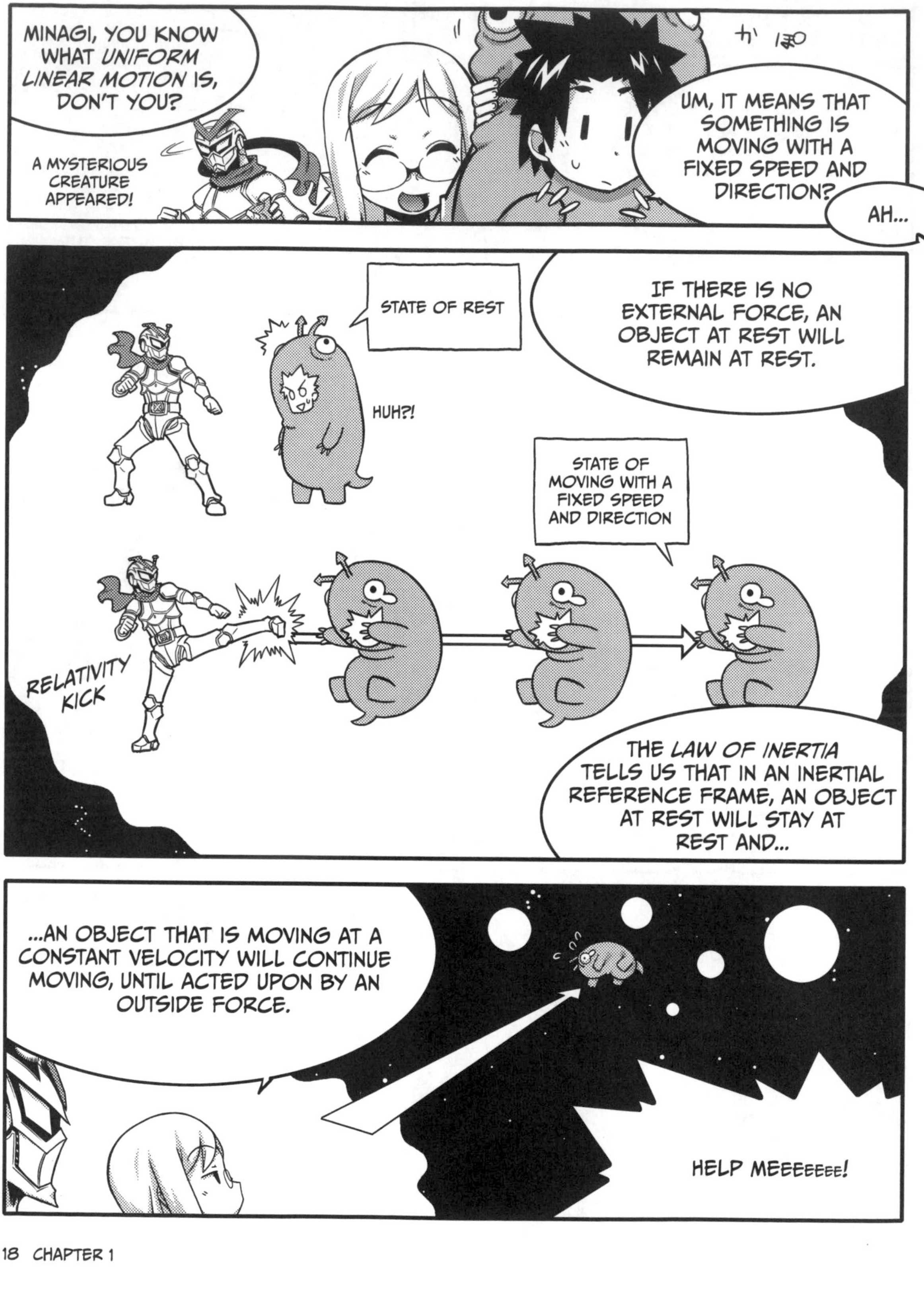
MINAGI, YOU KNOW WHAT UNIFORM LINEAR MOTION IS, DON'T YOU?
A MYSTERIOUS CREATURE APPEARED!
UM, IT MEANS THAT SOMETHING IS MOVING WITH A FIXED SPEED AND DIRECTION?
AH...
STATE OF REST
IF THERE IS NO EXTERNAL FORCE, AN OBJECT AT REST WILL REMAIN AT REST.
HUH?!
STATE OF MOVING WITH A FIXED SPEED AND DIRECTION
RELATIVITY KICK
THE LAW OF INERTIA TELLS US THAT IN AN INERTIAL REFERENCE FRAME, AN OBJECT AT REST WILL STAY AT REST AND...
...AN OBJECT THAT IS MOVING AT A CONSTANT VELOCITY WILL CONTINUE MOVING, UNTIL ACTED UPON BY AN OUTSIDE FORCE.
HELP MEEEEEEE!

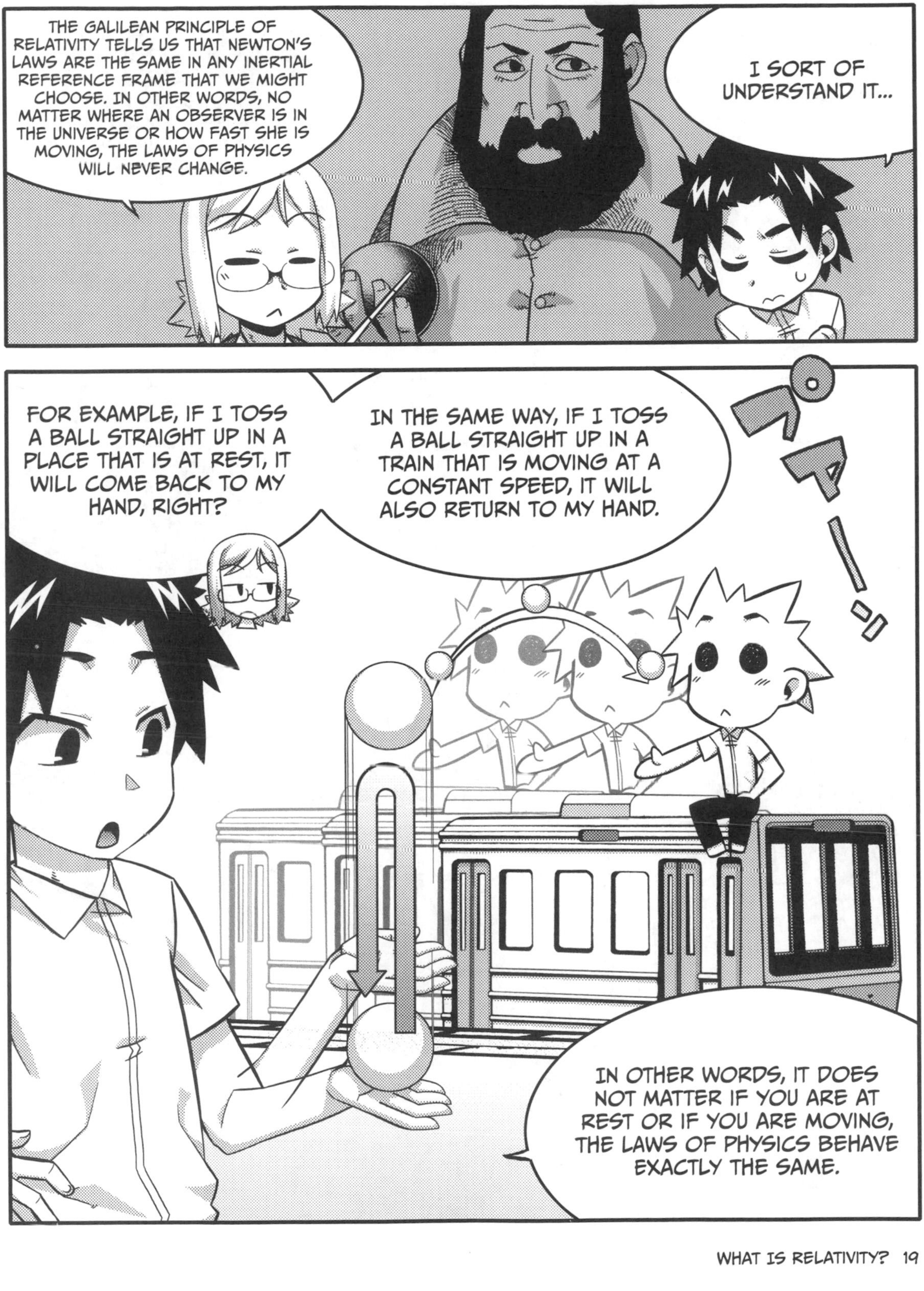
THE GALILEAN PRINCIPLE OF RELATIVITY TELLS US THAT NEWTON'S LAWS ARE THE SAME IN ANY INERTIAL REFERENCE FRAME THAT WE MIGHT CHOOSE. IN OTHER WORDS, NO MATTER WHERE AN OBSERVER IS IN THE UNIVERSE OR HOW FAST SHE IS MOVING, THE LAWS OF PHYSICS WILL NEVER CHANGE.
I SORT OF UNDERSTAND IT...
FOR EXAMPLE, IF I TOSS A BALL STRAIGHT UP IN A PLACE THAT IS AT REST, IT WILL COME BACK TO MY HAND, RIGHT?
IN THE SAME WAY, IF I TOSS A BALL STRAIGHT UP IN A TRAIN THAT IS MOVING AT A CONSTANT SPEED, IT WILL ALSO RETURN TO MY HAND.
IN OTHER WORDS, IT DOES NOT MATTER IF YOU ARE AT REST OR IF YOU ARE MOVING, THE LAWS OF PHYSICS BEHAVE EXACTLY THE SAME.

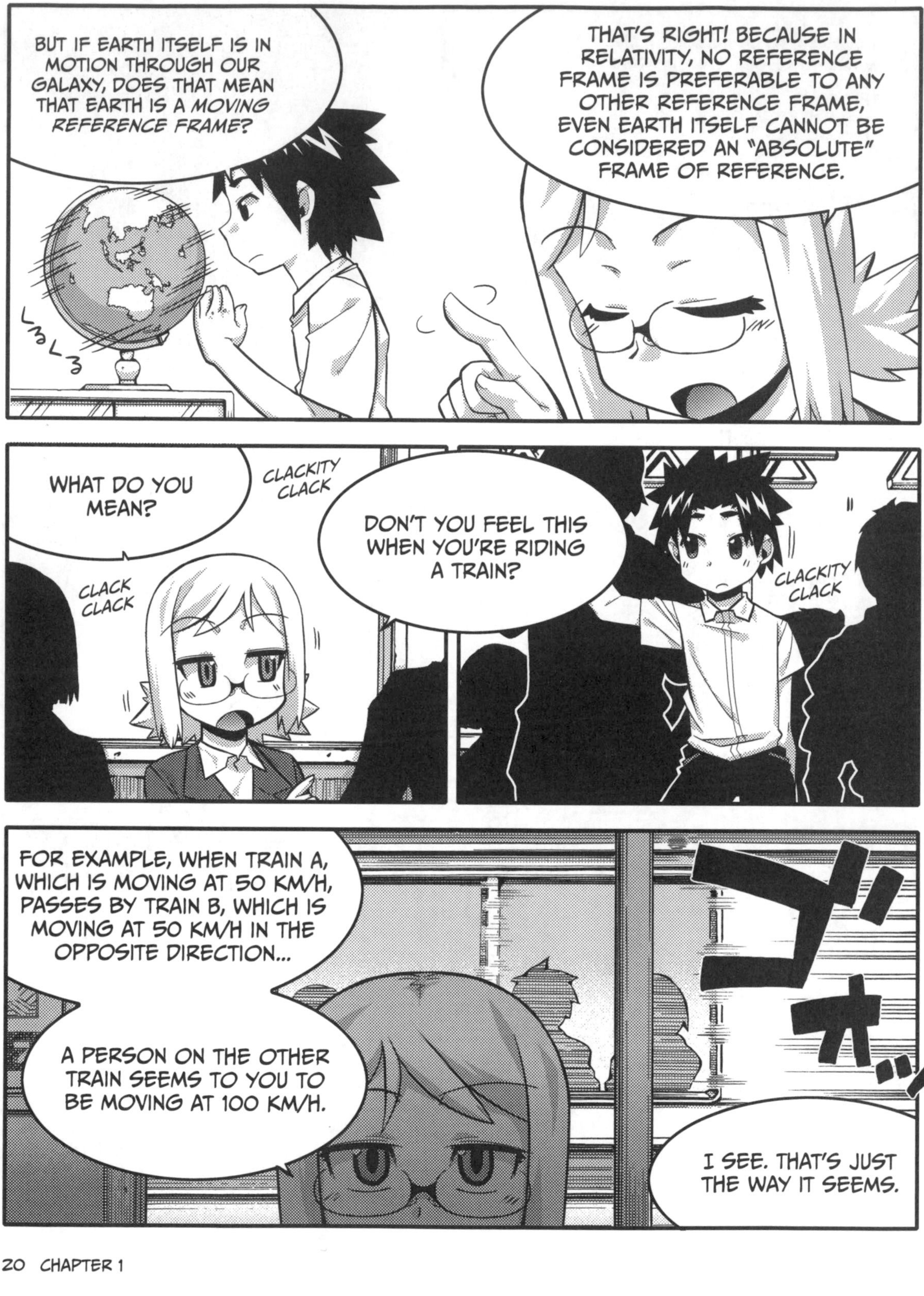
BUT IF EARTH ITSELF IS IN MOTION THROUGH OUR GALAXY, DOES THAT MEAN THAT EARTH IS A *MOVING REFERENCE FRAME*?
THAT'S RIGHT! BECAUSE IN RELATIVITY, NO REFERENCE FRAME IS PREFERABLE TO ANY OTHER REFERENCE FRAME, EVEN EARTH ITSELF CANNOT BE CONSIDERED AN "ABSOLUTE" FRAME OF REFERENCE.
WHAT DO YOU MEAN?
CLACKITY CLACK
CLACK CLACK
DON'T YOU FEEL THIS WHEN YOU'RE RIDING A TRAIN?
CLACKITY CLACK
FOR EXAMPLE, WHEN TRAIN A, WHICH IS MOVING AT 50 KM/H, PASSES BY TRAIN B, WHICH IS MOVING AT 50 KM/H IN THE OPPOSITE DIRECTION...
A PERSON ON THE OTHER TRAIN SEEMS TO YOU TO BE MOVING AT 100 KM/H.
I SEE. THAT'S JUST THE WAY IT SEEMS.

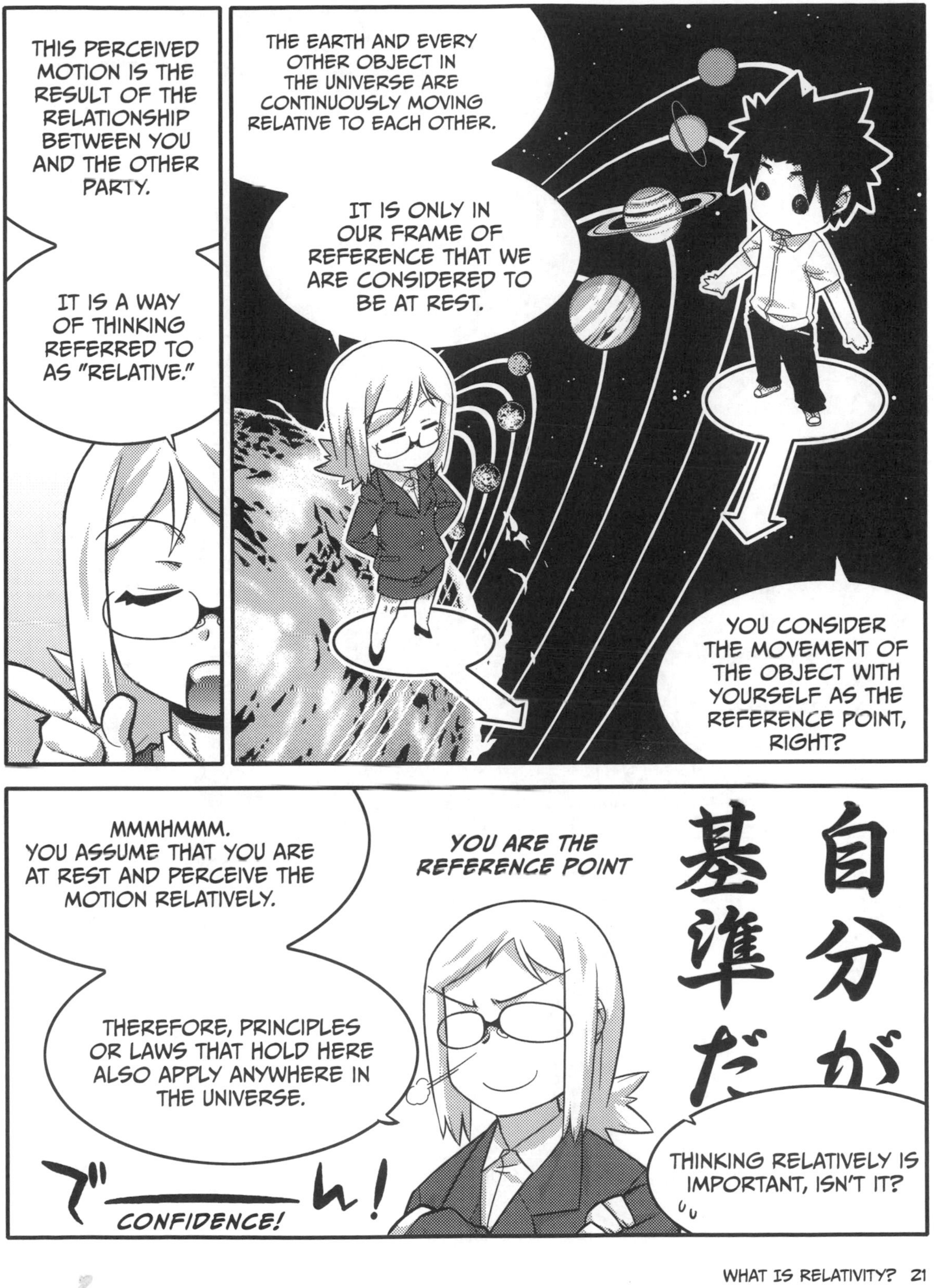
THIS PERCEIVED MOTION IS THE RESULT OF THE RELATIONSHIP BETWEEN YOU AND THE OTHER PARTY.
IT IS A WAY OF THINKING REFERRED TO AS "RELATIVE."
THE EARTH AND EVERY OTHER OBJECT IN THE UNIVERSE ARE CONTINUOUSLY MOVING RELATIVE TO EACH OTHER.
IT IS ONLY IN OUR FRAME OF REFERENCE THAT WE ARE CONSIDERED TO BE AT REST.
YOU CONSIDER THE MOVEMENT OF THE OBJECT WITH YOURSELF AS THE REFERENCE POINT, RIGHT?
MMMHMMM. YOU ASSUME THAT YOU ARE AT REST AND PERCEIVE THE MOTION RELATIVELY.
THEREFORE, PRINCIPLES OR LAWS THAT HOLD HERE ALSO APPLY ANYWHERE IN THE UNIVERSE.
自分が基準だ
YOU ARE THE REFERENCE POINT
でーーん!
CONFIDENCE!
THINKING RELATIVELY IS IMPORTANT, ISN'T IT?

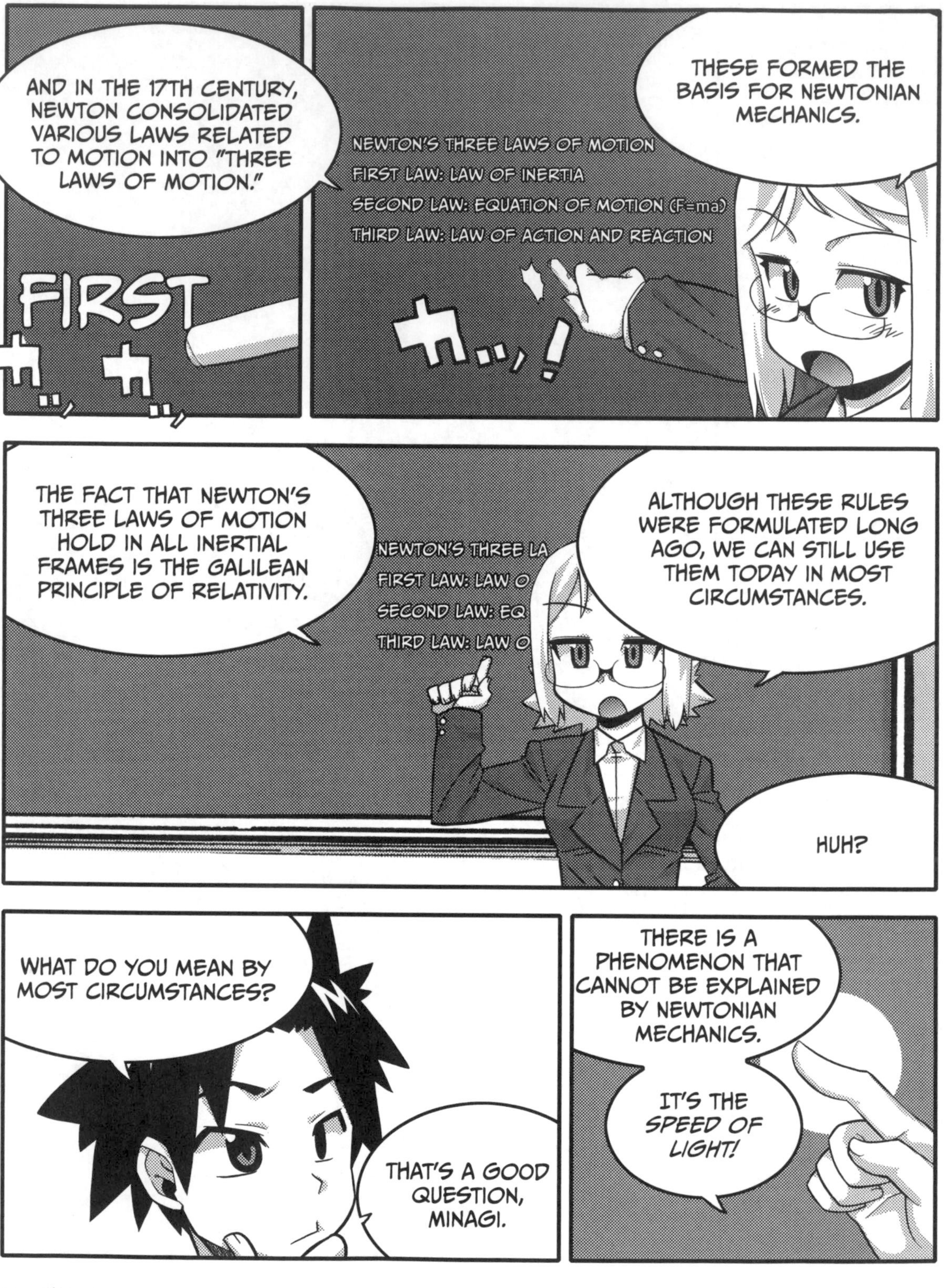

AND IN THE 17TH CENTURY, NEWTON CONSOLIDATED VARIOUS LAWS RELATED TO MOTION INTO "THREE LAWS OF MOTION."
FIRST
NEWTON'S THREE LAWS OF MOTION
FIRST LAW: LAW OF INERTIA
SECOND LAW: EQUATION OF MOTION (F=ma)
THIRD LAW: LAW OF ACTION AND REACTION
THESE FORMED THE BASIS FOR NEWTONIAN MECHANICS.
THE FACT THAT NEWTON'S THREE LAWS OF MOTION HOLD IN ALL INERTIAL FRAMES IS THE GALILEAN PRINCIPLE OF RELATIVITY.
ALTHOUGH THESE RULES WERE FORMULATED LONG AGO, WE CAN STILL USE THEM TODAY IN MOST CIRCUMSTANCES.
HUH?
WHAT DO YOU MEAN BY MOST CIRCUMSTANCES?
THAT'S A GOOD QUESTION, MINAGI.
THERE IS A PHENOMENON THAT CANNOT BE EXPLAINED BY NEWTONIAN MECHANICS.
IT'S THE *SPEED OF LIGHT!*

3. MYSTERY OF THE SPEED OF LIGHT

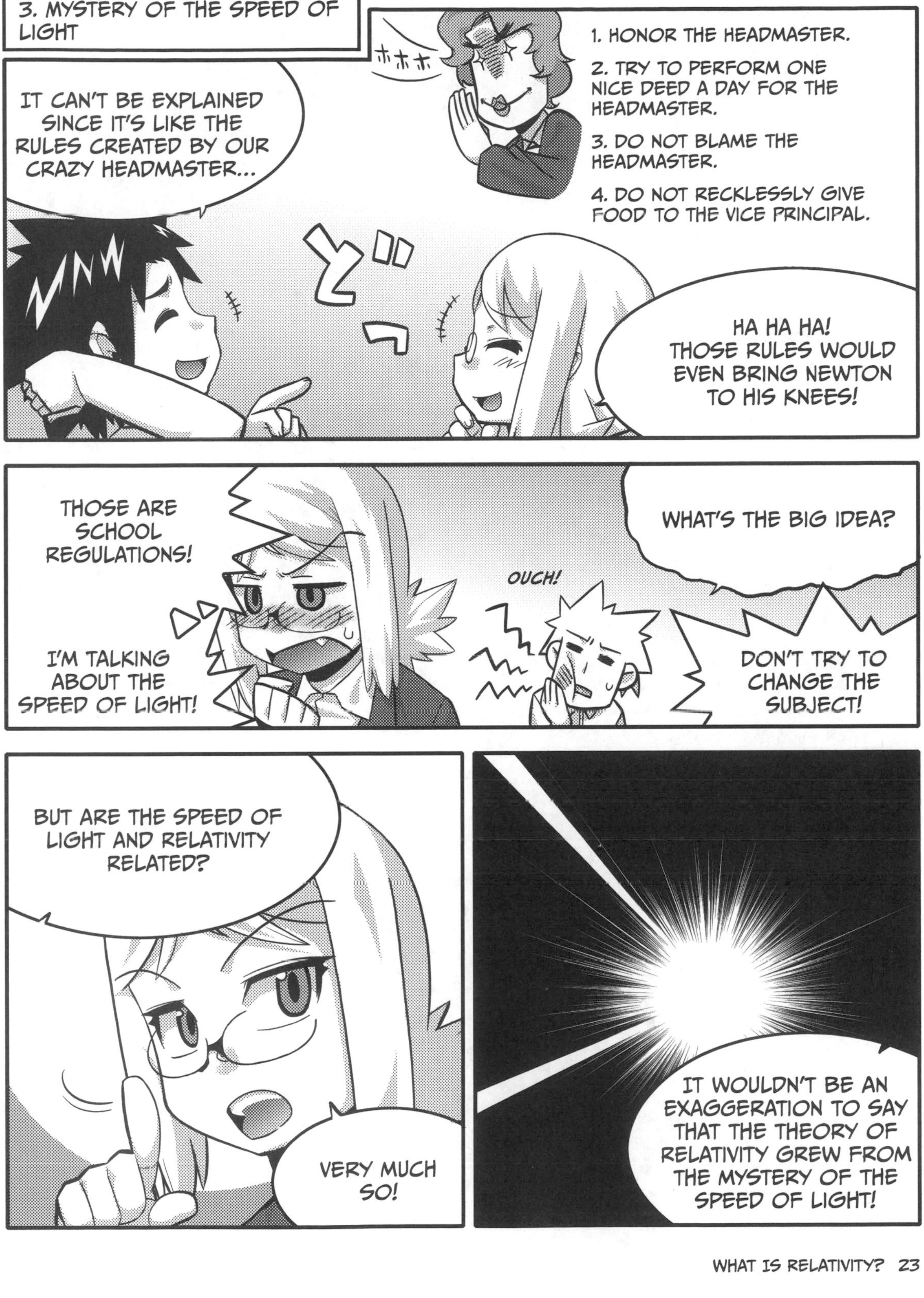
IT CAN'T BE EXPLAINED SINCE IT'S LIKE THE RULES CREATED BY OUR CRAZY HEADMASTER...
ホホホ
1. HONOR THE HEADMASTER.
2. TRY TO PERFORM ONE NICE DEED A DAY FOR THE HEADMASTER.
3. DO NOT BLAME THE HEADMASTER.
4. DO NOT RECKLESSLY GIVE FOOD TO THE VICE PRINCIPAL.
どっ
HA HA HA! THOSE RULES WOULD EVEN BRING NEWTON TO HIS KNEES!
THOSE ARE SCHOOL REGULATIONS!
I'M TALKING ABOUT THE SPEED OF LIGHT!
OUCH!
WHAT'S THE BIG IDEA?
DON'T TRY TO CHANGE THE SUBJECT!
BUT ARE THE SPEED OF LIGHT AND RELATIVITY RELATED?
VERY MUCH SO!
IT WOULDN'T BE AN EXAGGERATION TO SAY THAT THE THEORY OF RELATIVITY GREW FROM THE MYSTERY OF THE SPEED OF LIGHT!

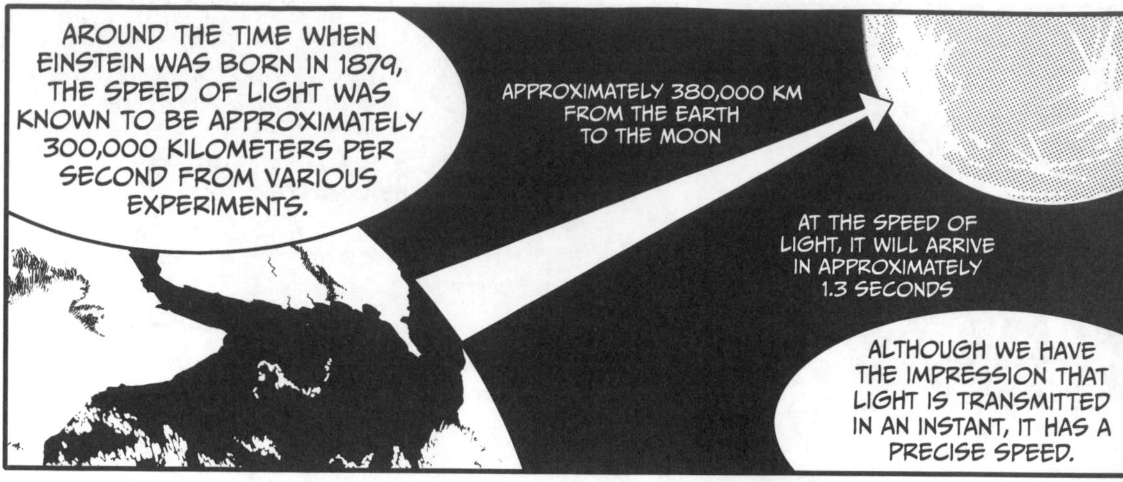
AROUND THE TIME WHEN EINSTEIN WAS BORN IN 1879, THE SPEED OF LIGHT WAS KNOWN TO BE APPROXIMATELY 300,000 KILOMETERS PER SECOND FROM VARIOUS EXPERIMENTS.
APPROXIMATELY 380,000 KM FROM THE EARTH TO THE MOON
AT THE SPEED OF LIGHT, IT WILL ARRIVE IN APPROXIMATELY 1.3 SECONDS
ALTHOUGH WE HAVE THE IMPRESSION THAT LIGHT IS TRANSMITTED IN AN INSTANT, IT HAS A PRECISE SPEED.

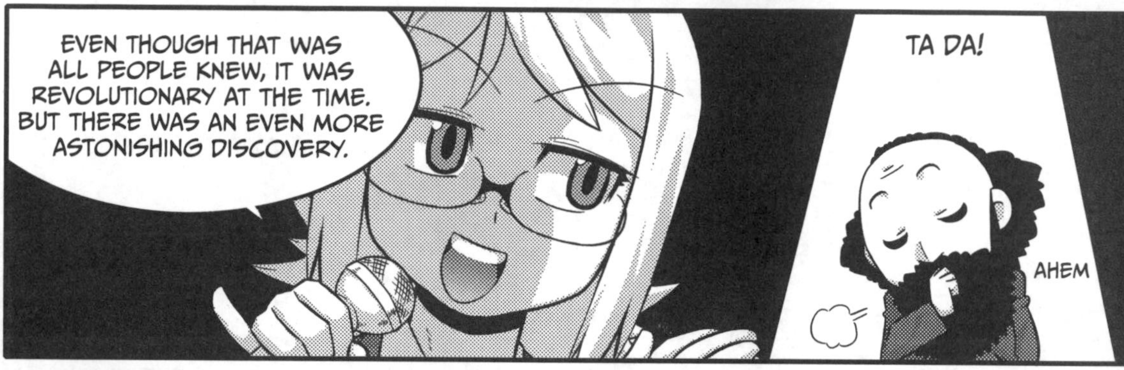
EVEN THOUGH THAT WAS ALL PEOPLE KNEW, IT WAS REVOLUTIONARY AT THE TIME. BUT THERE WAS AN EVEN MORE ASTONISHING DISCOVERY.
TA DA!
AHEM

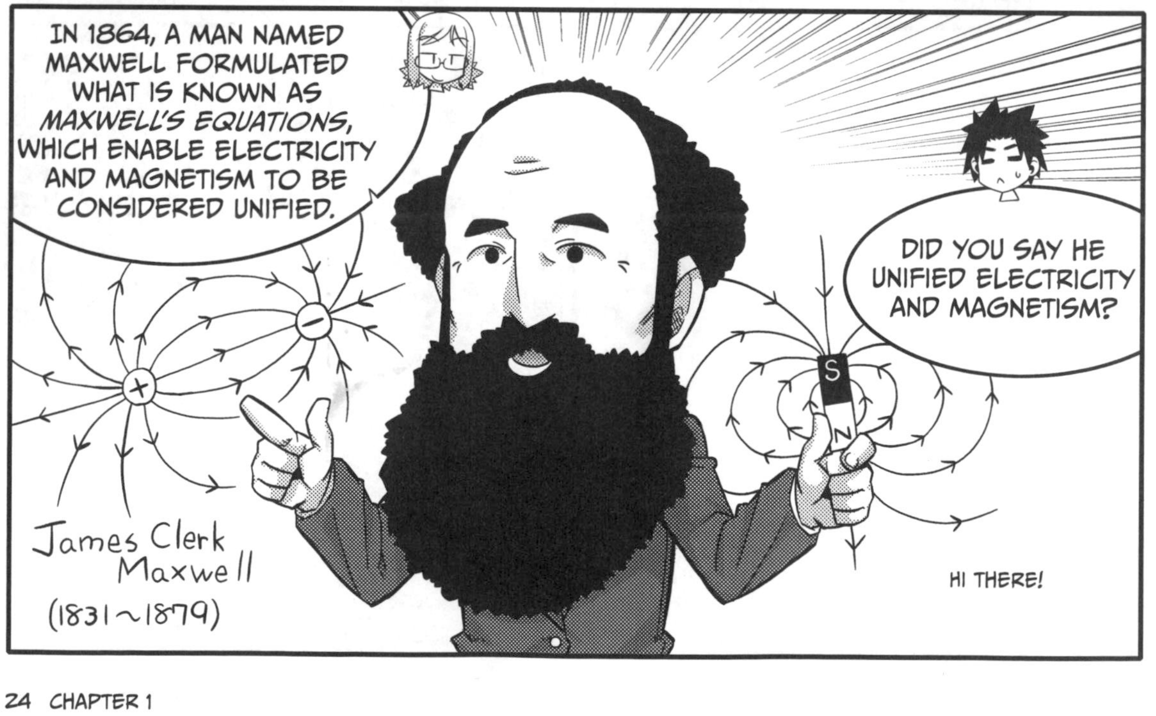
IN 1864, A MAN NAMED MAXWELL FORMULATED WHAT IS KNOWN AS *MAXWELL'S EQUATIONS*, WHICH ENABLE ELECTRICITY AND MAGNETISM TO BE CONSIDERED UNIFIED.
DID YOU SAY HE UNIFIED ELECTRICITY AND MAGNETISM?
+
−
S
N
James Clerk Maxwell (1831~1879)
HI THERE!

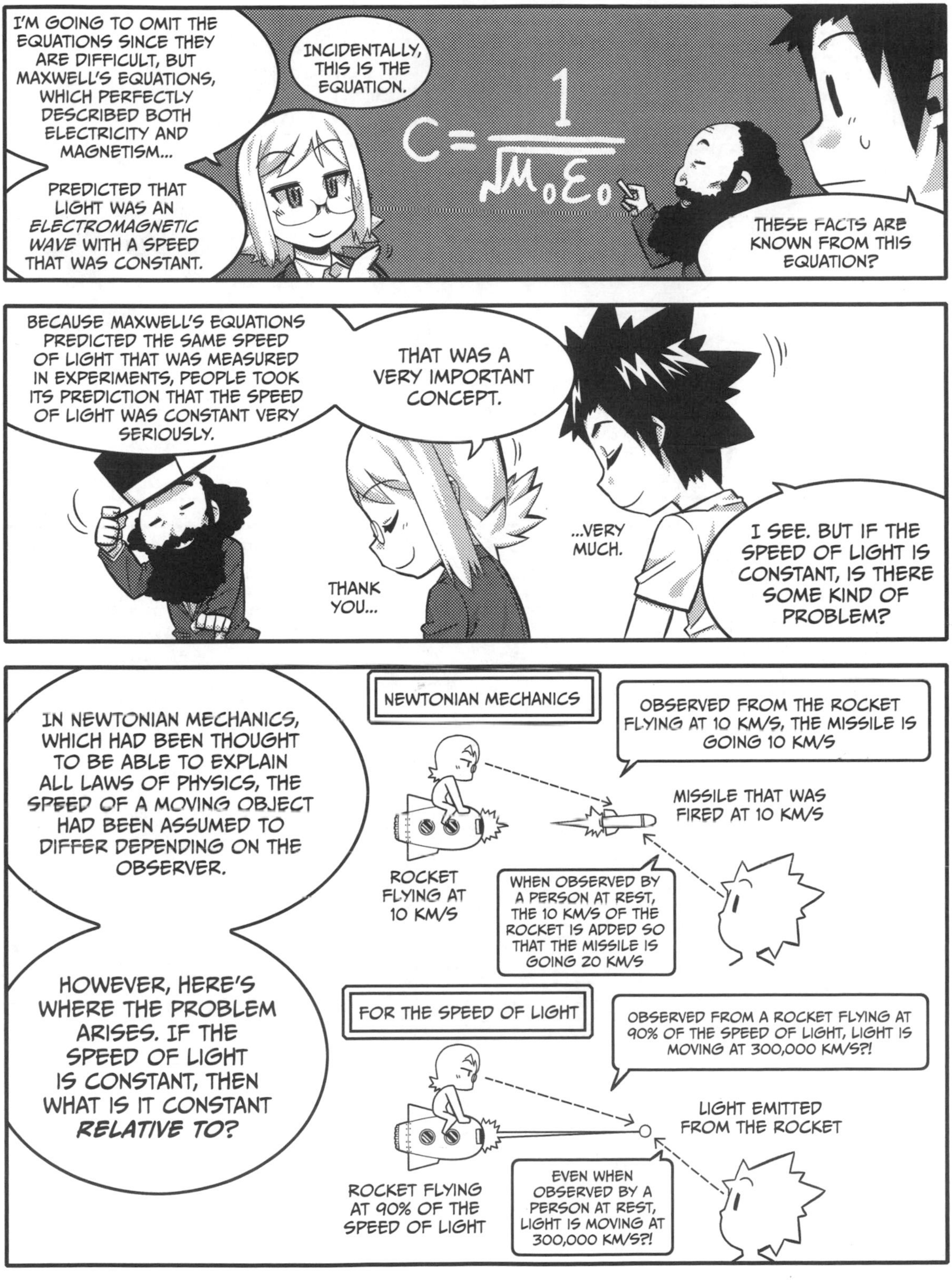
I'M GOING TO OMIT THE EQUATIONS SINCE THEY ARE DIFFICULT, BUT MAXWELL'S EQUATIONS, WHICH PERFECTLY DESCRIBED BOTH ELECTRICITY AND MAGNETISM...
PREDICTED THAT LIGHT WAS AN *ELECTROMAGNETIC WAVE* WITH A SPEED THAT WAS CONSTANT.
INCIDENTALLY, THIS IS THE EQUATION.
$C = \frac{1}{\sqrt{\mu_0 \varepsilon_0}}$
THESE FACTS ARE KNOWN FROM THIS EQUATION?
BECAUSE MAXWELL'S EQUATIONS PREDICTED THE SAME SPEED OF LIGHT THAT WAS MEASURED IN EXPERIMENTS, PEOPLE TOOK ITS PREDICTION THAT THE SPEED OF LIGHT WAS CONSTANT VERY SERIOUSLY.
THAT WAS A VERY IMPORTANT CONCEPT.
THANK YOU...
...VERY MUCH.
I SEE. BUT IF THE SPEED OF LIGHT IS CONSTANT, IS THERE SOME KIND OF PROBLEM?
IN NEWTONIAN MECHANICS, WHICH HAD BEEN THOUGHT TO BE ABLE TO EXPLAIN ALL LAWS OF PHYSICS, THE SPEED OF A MOVING OBJECT HAD BEEN ASSUMED TO DIFFER DEPENDING ON THE OBSERVER.
NEWTONIAN MECHANICS
OBSERVED FROM THE ROCKET FLYING AT 10 KM/S, THE MISSILE IS GOING 10 KM/S
MISSILE THAT WAS FIRED AT 10 KM/S
ROCKET FLYING AT 10 KM/S
WHEN OBSERVED BY A PERSON AT REST, THE 10 KM/S OF THE ROCKET IS ADDED SO THAT THE MISSILE IS GOING 20 KM/S
HOWEVER, HERE'S WHERE THE PROBLEM ARISES. IF THE SPEED OF LIGHT IS CONSTANT, THEN WHAT IS IT CONSTANT ***RELATIVE TO?***
FOR THE SPEED OF LIGHT
OBSERVED FROM A ROCKET FLYING AT 90% OF THE SPEED OF LIGHT, LIGHT IS MOVING AT 300,000 KM/S?!
LIGHT EMITTED FROM THE ROCKET
ROCKET FLYING AT 90% OF THE SPEED OF LIGHT
EVEN WHEN OBSERVED BY A PERSON AT REST, LIGHT IS MOVING AT 300,000 KM/S?!

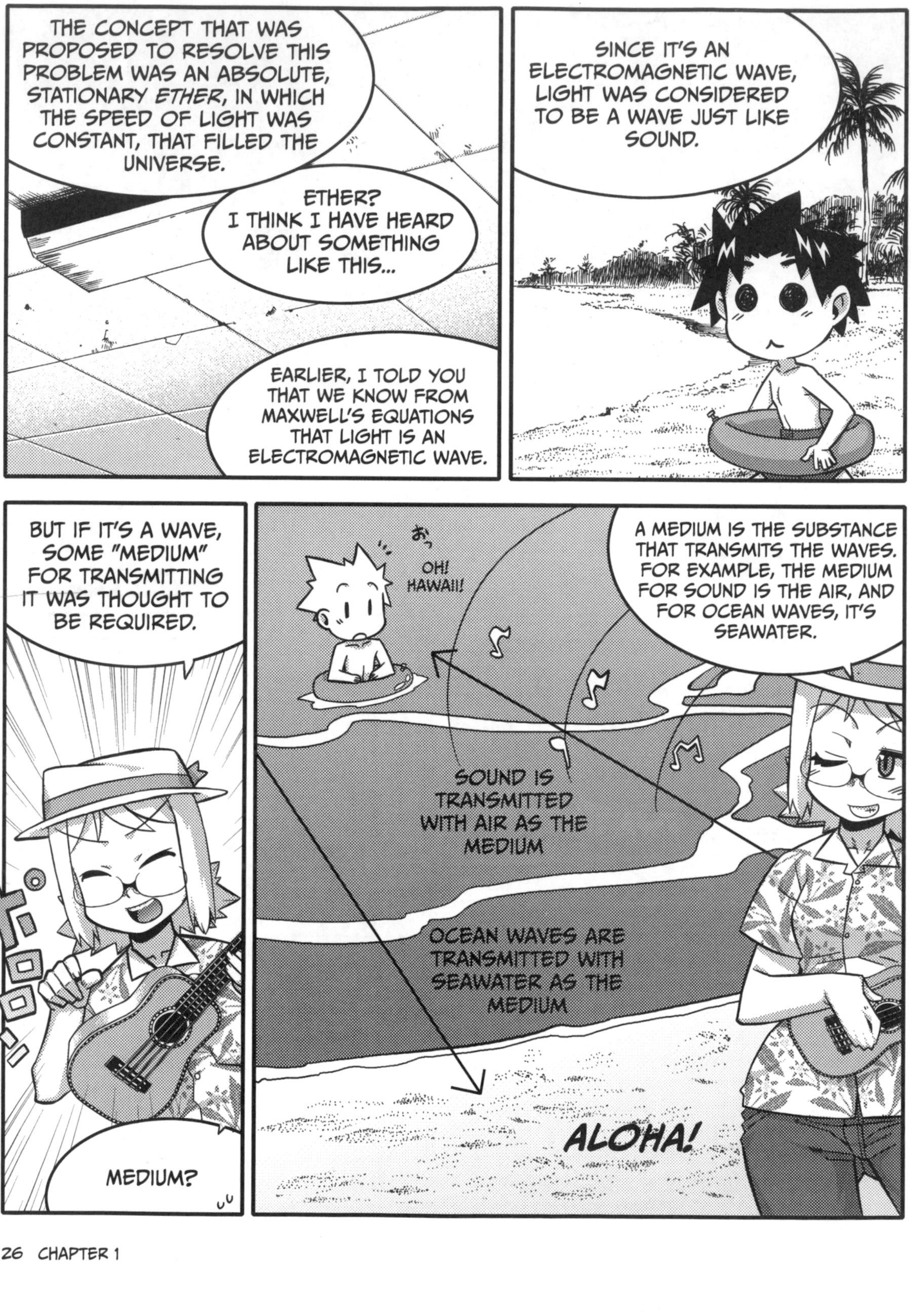

THE CONCEPT THAT WAS PROPOSED TO RESOLVE THIS PROBLEM WAS AN ABSOLUTE, STATIONARY *ETHER*, IN WHICH THE SPEED OF LIGHT WAS CONSTANT, THAT FILLED THE UNIVERSE.
ETHER? I THINK I HAVE HEARD ABOUT SOMETHING LIKE THIS...
EARLIER, I TOLD YOU THAT WE KNOW FROM MAXWELL'S EQUATIONS THAT LIGHT IS AN ELECTROMAGNETIC WAVE.
SINCE IT'S AN ELECTROMAGNETIC WAVE, LIGHT WAS CONSIDERED TO BE A WAVE JUST LIKE SOUND.
BUT IF IT'S A WAVE, SOME "MEDIUM" FOR TRANSMITTING IT WAS THOUGHT TO BE REQUIRED.
ポロロン
MEDIUM?
おっ
OH! HAWAII!
A MEDIUM IS THE SUBSTANCE THAT TRANSMITS THE WAVES. FOR EXAMPLE, THE MEDIUM FOR SOUND IS THE AIR, AND FOR OCEAN WAVES, IT'S SEAWATER.
SOUND IS TRANSMITTED WITH AIR AS THE MEDIUM
OCEAN WAVES ARE TRANSMITTED WITH SEAWATER AS THE MEDIUM
ALOHA!

OF COURSE, SINCE THERE CERTAINLY IS NO AIR IN SPACE, SOUND IS NOT TRANSMITTED, BUT...

...LIGHT SOMEHOW REACHES EARTH.

WELL, THEN IT'S NATURAL TO THINK THAT SPACE CONTAINS A MEDIUM FOR TRANSMITTING LIGHT!

THEREFORE, SCIENTISTS PROPOSED THE IDEA THAT AN UNKNOWN MEDIUM *ETHER* FILLED THE VACUUM OF SPACE.

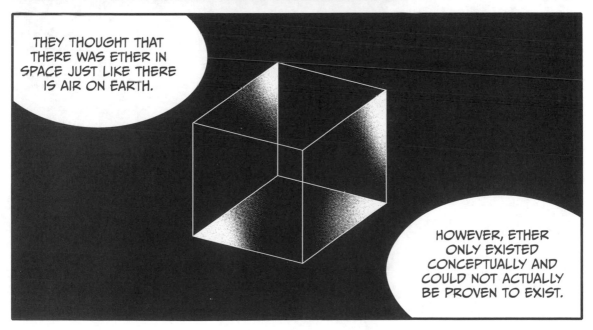

THEY THOUGHT THAT THERE WAS ETHER IN SPACE JUST LIKE THERE IS AIR ON EARTH.

HOWEVER, ETHER ONLY EXISTED CONCEPTUALLY AND COULD NOT ACTUALLY BE PROVEN TO EXIST.

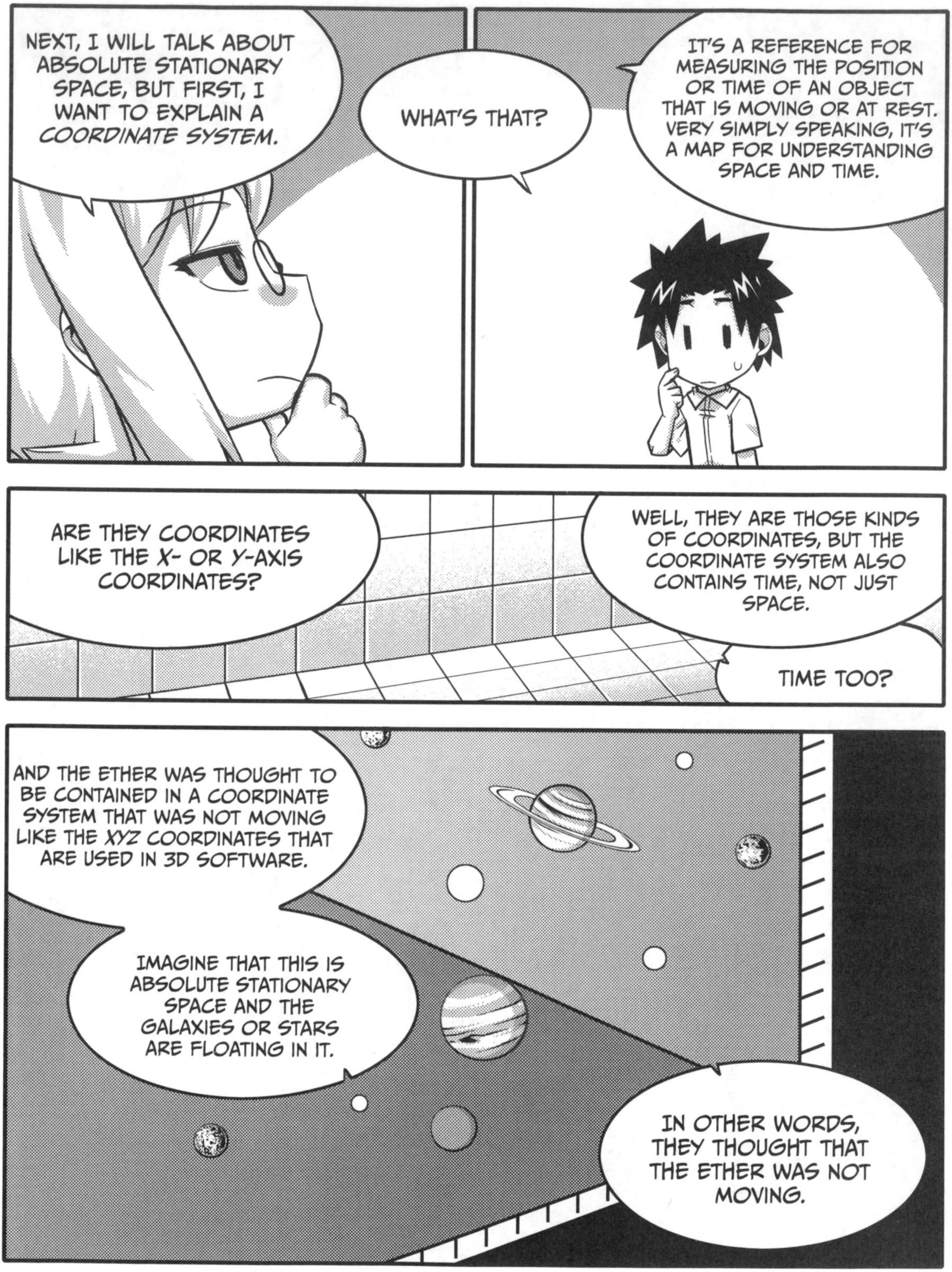

NEXT, I WILL TALK ABOUT ABSOLUTE STATIONARY SPACE, BUT FIRST, I WANT TO EXPLAIN A *COORDINATE SYSTEM.*
WHAT'S THAT?
IT'S A REFERENCE FOR MEASURING THE POSITION OR TIME OF AN OBJECT THAT IS MOVING OR AT REST. VERY SIMPLY SPEAKING, IT'S A MAP FOR UNDERSTANDING SPACE AND TIME.
ARE THEY COORDINATES LIKE THE *X*- OR *Y*-AXIS COORDINATES?
WELL, THEY ARE THOSE KINDS OF COORDINATES, BUT THE COORDINATE SYSTEM ALSO CONTAINS TIME, NOT JUST SPACE.
TIME TOO?
AND THE ETHER WAS THOUGHT TO BE CONTAINED IN A COORDINATE SYSTEM THAT WAS NOT MOVING LIKE THE *XYZ* COORDINATES THAT ARE USED IN 3D SOFTWARE.
IMAGINE THAT THIS IS ABSOLUTE STATIONARY SPACE AND THE GALAXIES OR STARS ARE FLOATING IN IT.
IN OTHER WORDS, THEY THOUGHT THAT THE ETHER WAS NOT MOVING.

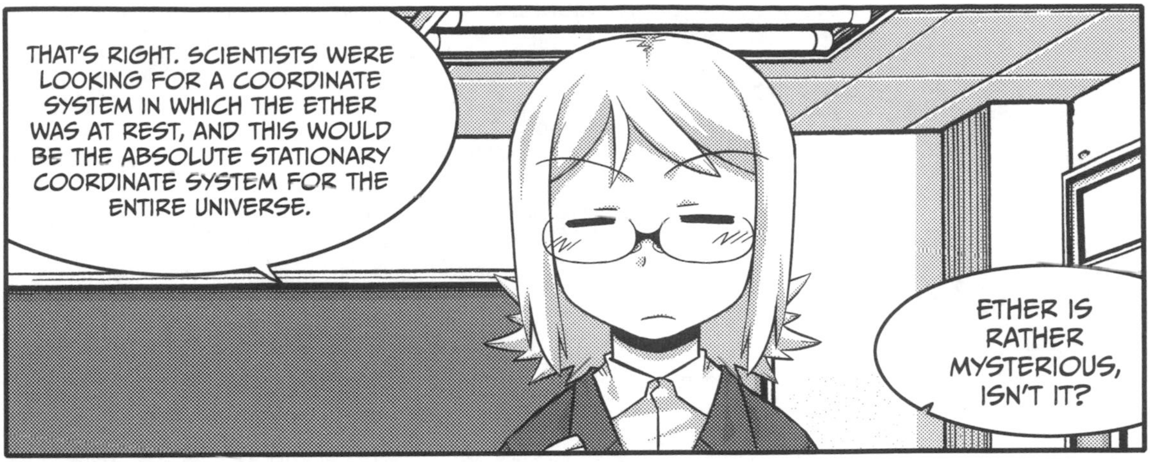
THAT'S RIGHT. SCIENTISTS WERE LOOKING FOR A COORDINATE SYSTEM IN WHICH THE ETHER WAS AT REST, AND THIS WOULD BE THE ABSOLUTE STATIONARY COORDINATE SYSTEM FOR THE ENTIRE UNIVERSE.
ETHER IS RATHER MYSTERIOUS, ISN'T IT?

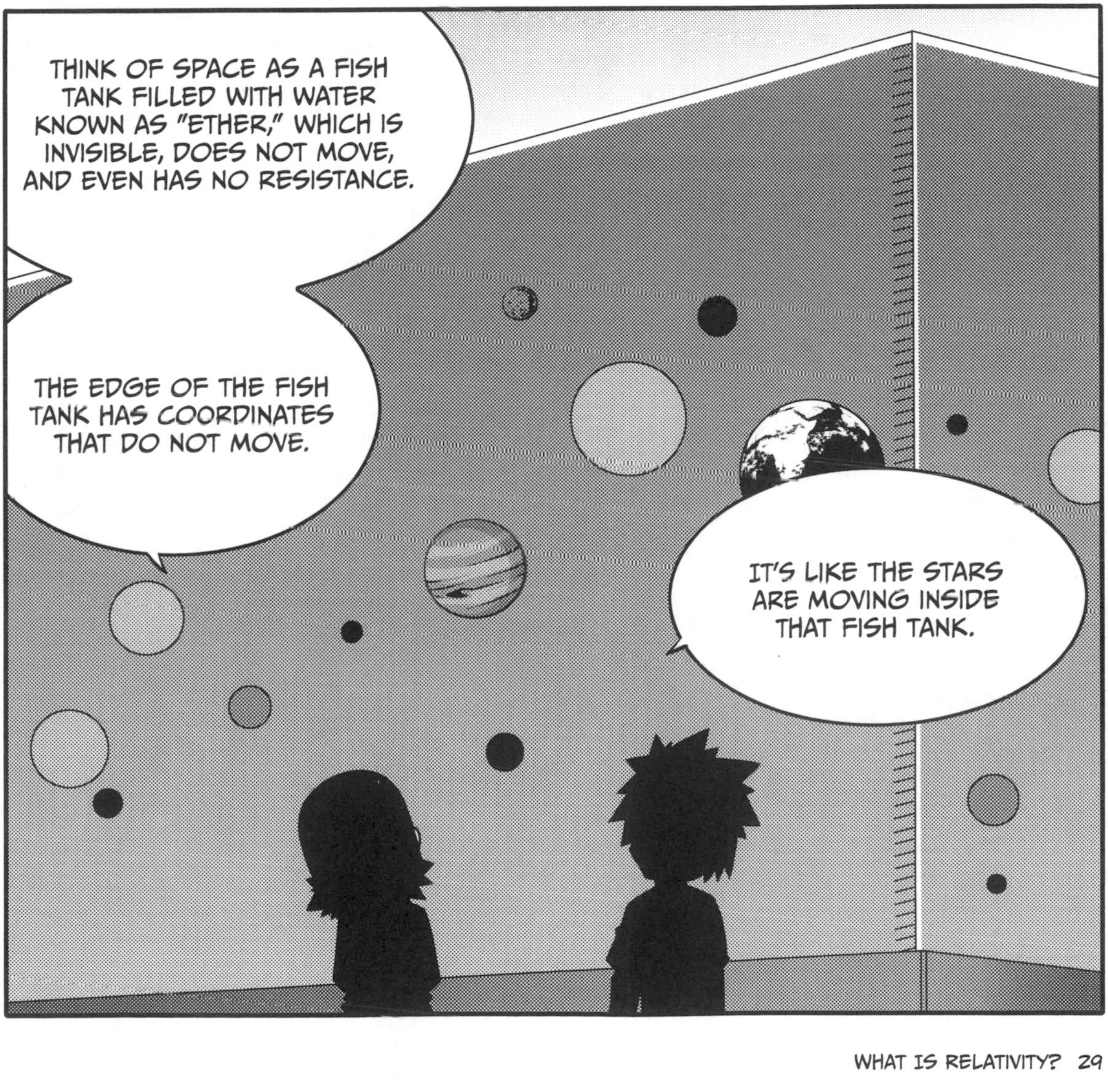
THINK OF SPACE AS A FISH TANK FILLED WITH WATER KNOWN AS "ETHER," WHICH IS INVISIBLE, DOES NOT MOVE, AND EVEN HAS NO RESISTANCE.
THE EDGE OF THE FISH TANK HAS COORDINATES THAT DO NOT MOVE.
IT'S LIKE THE STARS ARE MOVING INSIDE THAT FISH TANK.

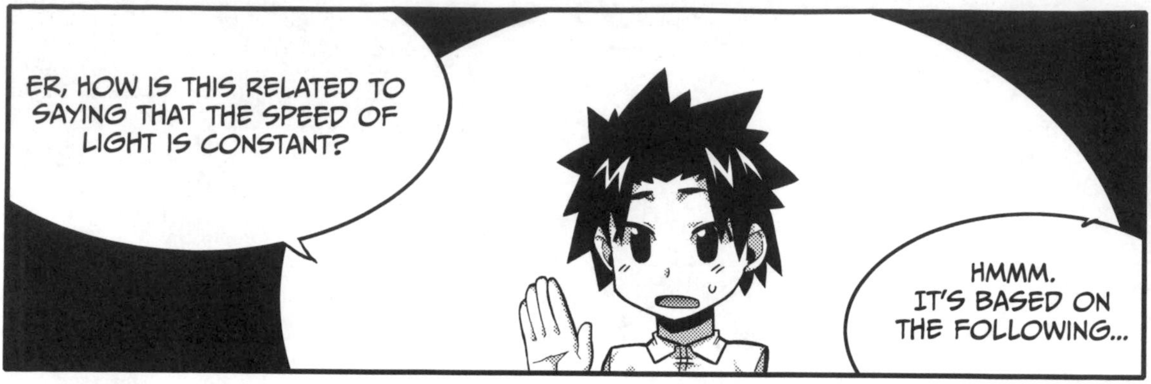
ER, HOW IS THIS RELATED TO SAYING THAT THE SPEED OF LIGHT IS CONSTANT?
HMMM. IT'S BASED ON THE FOLLOWING...

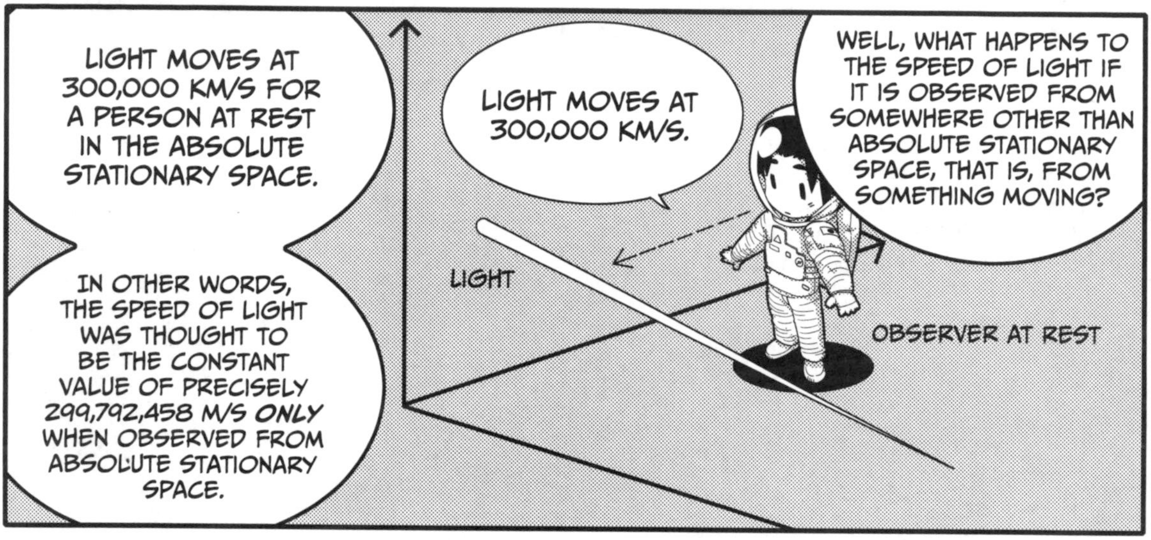
LIGHT MOVES AT 300,000 KM/S FOR A PERSON AT REST IN THE ABSOLUTE STATIONARY SPACE.
IN OTHER WORDS, THE SPEED OF LIGHT WAS THOUGHT TO BE THE CONSTANT VALUE OF PRECISELY 299,792,458 M/S *ONLY* WHEN OBSERVED FROM ABSOLUTE STATIONARY SPACE.
LIGHT MOVES AT 300,000 KM/S.
LIGHT
WELL, WHAT HAPPENS TO THE SPEED OF LIGHT IF IT IS OBSERVED FROM SOMEWHERE OTHER THAN ABSOLUTE STATIONARY SPACE, THAT IS, FROM SOMETHING MOVING?
OBSERVER AT REST

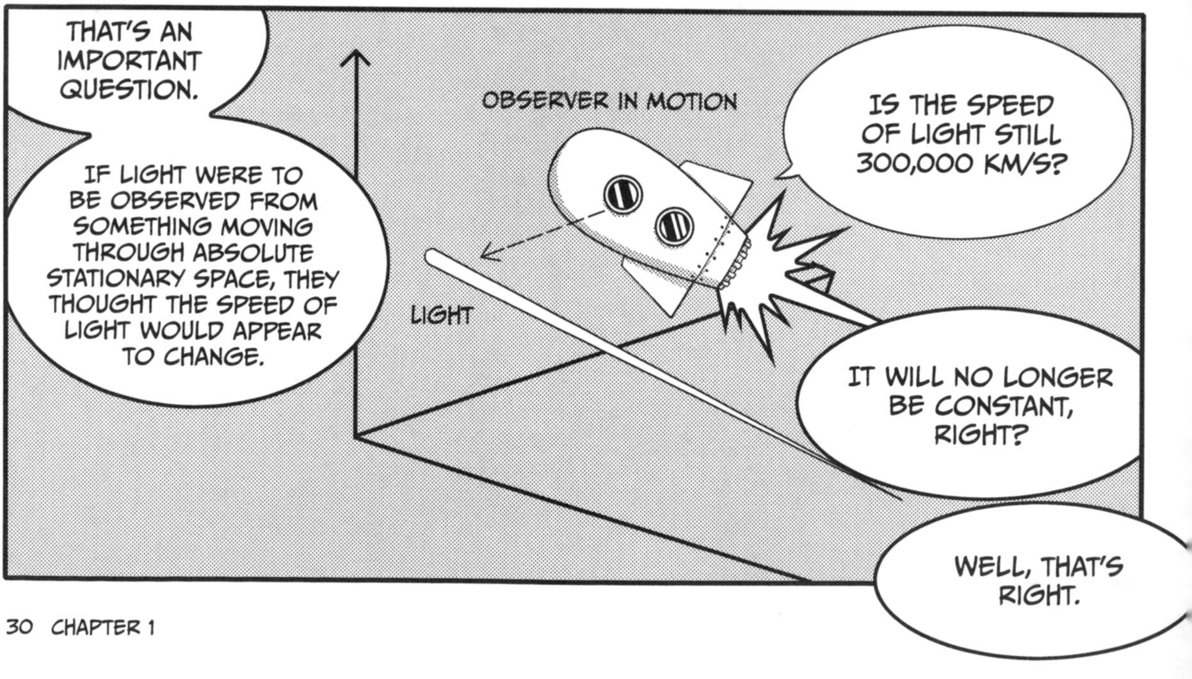
THAT'S AN IMPORTANT QUESTION.
IF LIGHT WERE TO BE OBSERVED FROM SOMETHING MOVING THROUGH ABSOLUTE STATIONARY SPACE, THEY THOUGHT THE SPEED OF LIGHT WOULD APPEAR TO CHANGE.
OBSERVER IN MOTION
LIGHT
IS THE SPEED OF LIGHT STILL 300,000 KM/S?
IT WILL NO LONGER BE CONSTANT, RIGHT?
WELL, THAT'S RIGHT.

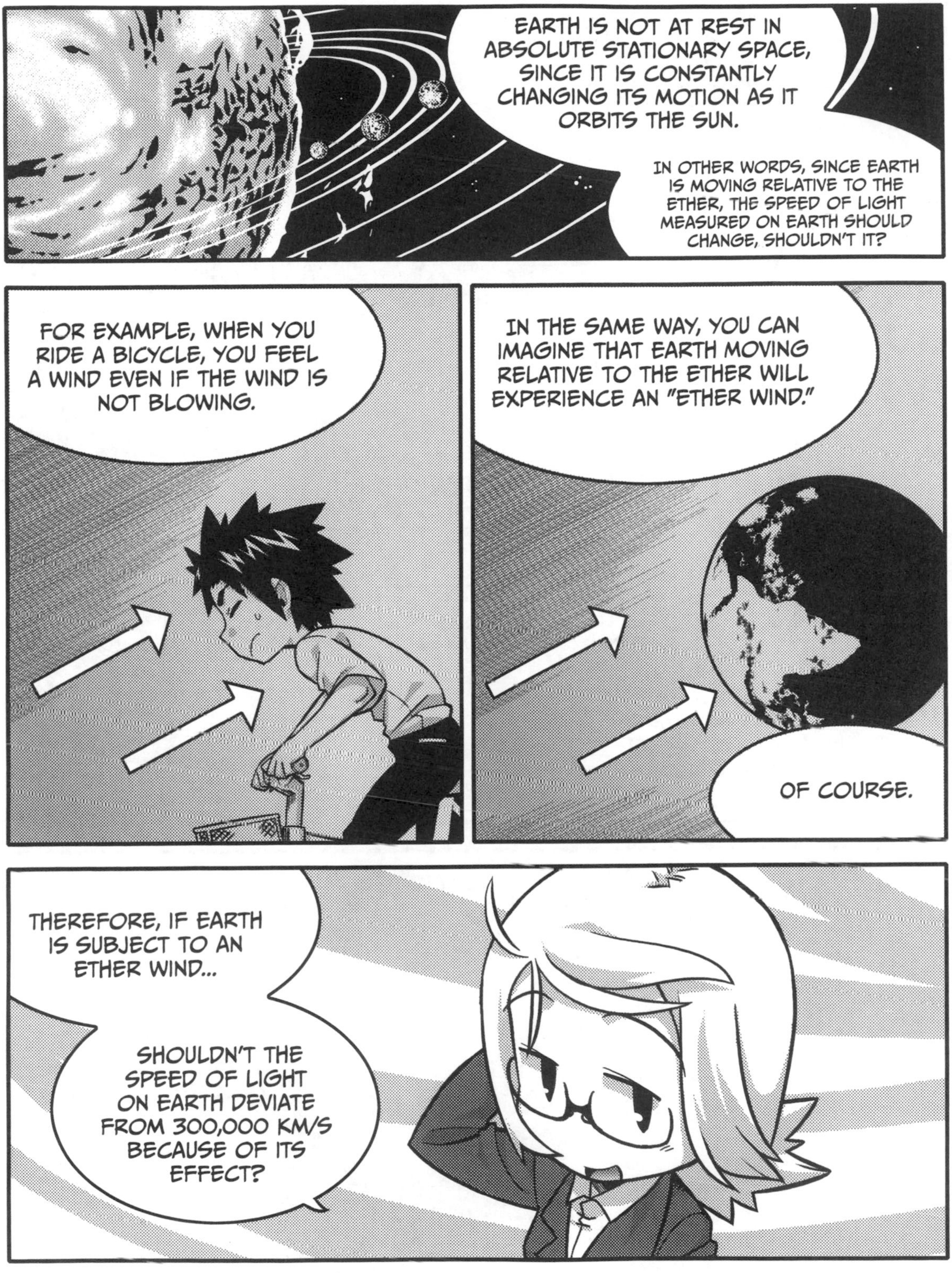
EARTH IS NOT AT REST IN ABSOLUTE STATIONARY SPACE, SINCE IT IS CONSTANTLY CHANGING ITS MOTION AS IT ORBITS THE SUN.
IN OTHER WORDS, SINCE EARTH IS MOVING RELATIVE TO THE ETHER, THE SPEED OF LIGHT MEASURED ON EARTH SHOULD CHANGE, SHOULDN'T IT?
FOR EXAMPLE, WHEN YOU RIDE A BICYCLE, YOU FEEL A WIND EVEN IF THE WIND IS NOT BLOWING.
IN THE SAME WAY, YOU CAN IMAGINE THAT EARTH MOVING RELATIVE TO THE ETHER WILL EXPERIENCE AN "ETHER WIND."
OF COURSE.
THEREFORE, IF EARTH IS SUBJECT TO AN ETHER WIND...
SHOULDN'T THE SPEED OF LIGHT ON EARTH DEVIATE FROM 300,000 KM/S BECAUSE OF ITS EFFECT?

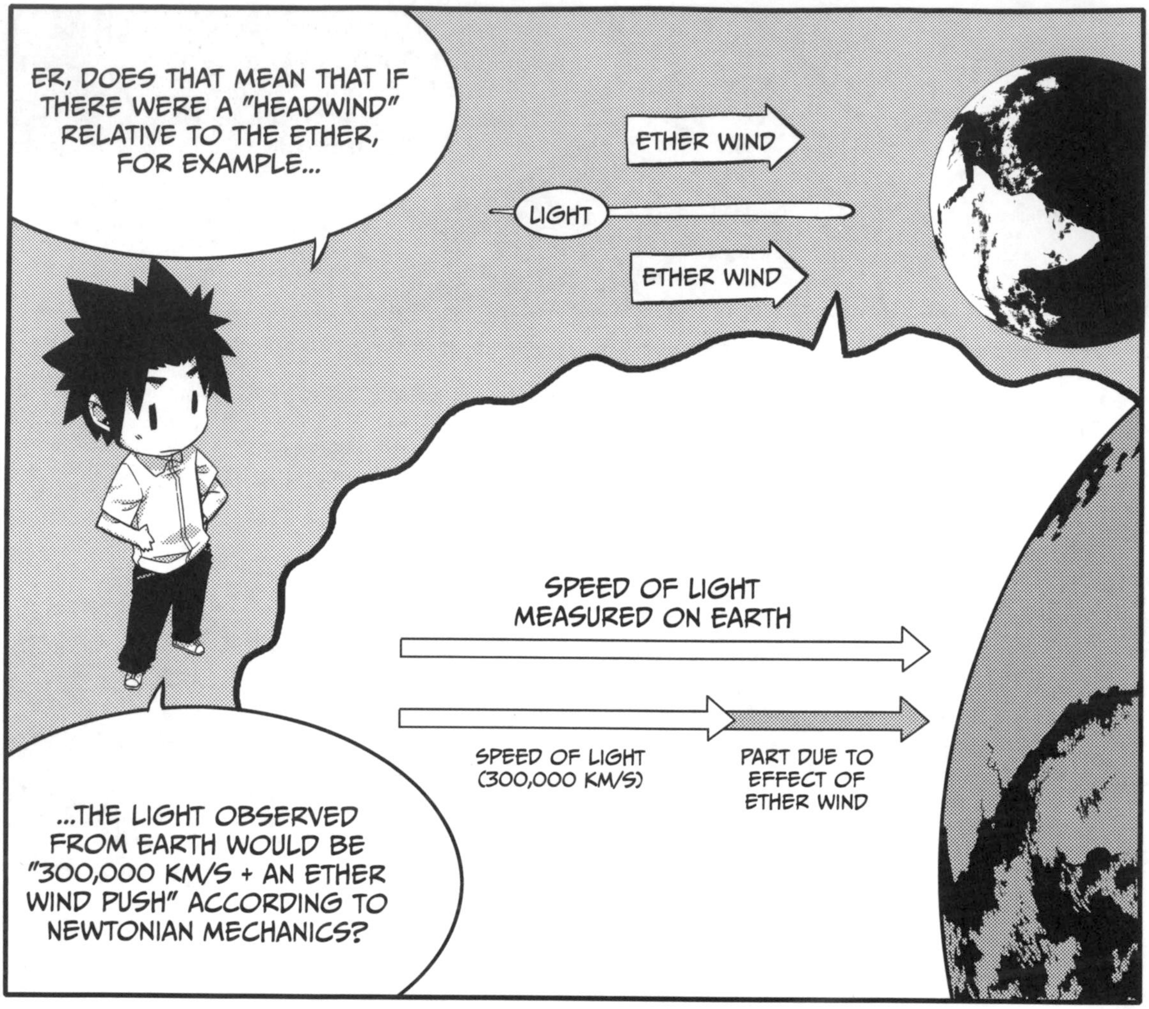
ER, DOES THAT MEAN THAT IF THERE WERE A "HEADWIND" RELATIVE TO THE ETHER, FOR EXAMPLE...
ETHER WIND
LIGHT
ETHER WIND
SPEED OF LIGHT MEASURED ON EARTH
SPEED OF LIGHT (300,000 KM/S)
PART DUE TO EFFECT OF ETHER WIND
...THE LIGHT OBSERVED FROM EARTH WOULD BE "300,000 KM/S + AN ETHER WIND PUSH" ACCORDING TO NEWTONIAN MECHANICS?

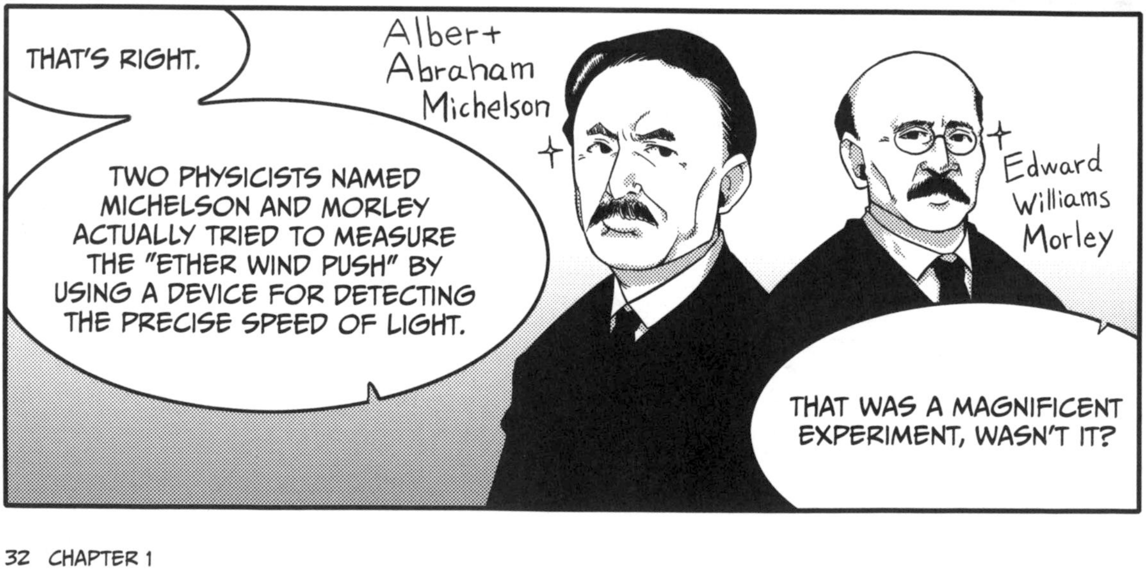
THAT'S RIGHT.
Albert Abraham Michelson
Edward Williams Morley
TWO PHYSICISTS NAMED MICHELSON AND MORLEY ACTUALLY TRIED TO MEASURE THE "ETHER WIND PUSH" BY USING A DEVICE FOR DETECTING THE PRECISE SPEED OF LIGHT.
THAT WAS A MAGNIFICENT EXPERIMENT, WASN'T IT?

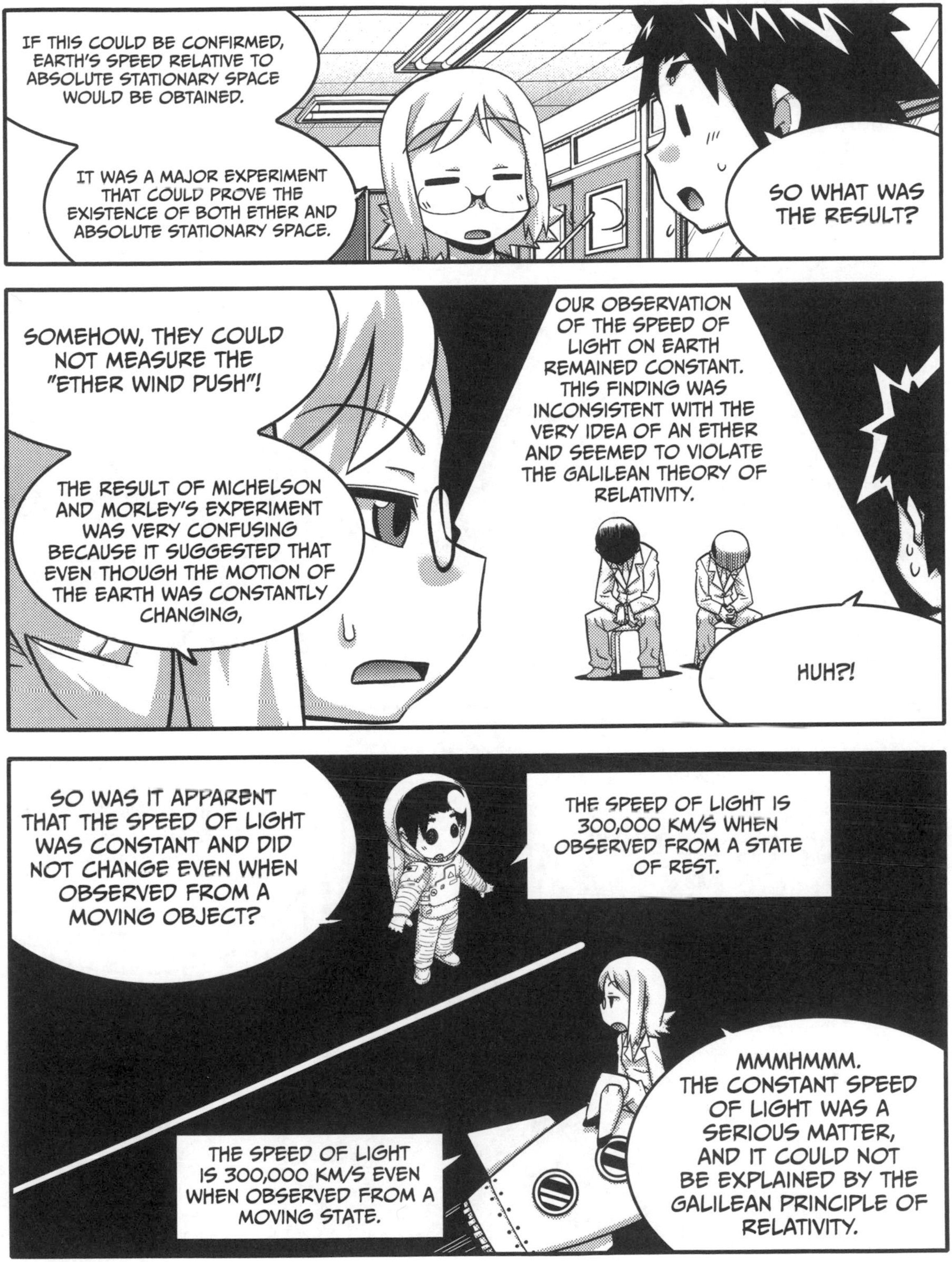
IF THIS COULD BE CONFIRMED, EARTH'S SPEED RELATIVE TO ABSOLUTE STATIONARY SPACE WOULD BE OBTAINED.
IT WAS A MAJOR EXPERIMENT THAT COULD PROVE THE EXISTENCE OF BOTH ETHER AND ABSOLUTE STATIONARY SPACE.
SO WHAT WAS THE RESULT?
SOMEHOW, THEY COULD NOT MEASURE THE "ETHER WIND PUSH"!
THE RESULT OF MICHELSON AND MORLEY'S EXPERIMENT WAS VERY CONFUSING BECAUSE IT SUGGESTED THAT EVEN THOUGH THE MOTION OF THE EARTH WAS CONSTANTLY CHANGING,
OUR OBSERVATION OF THE SPEED OF LIGHT ON EARTH REMAINED CONSTANT. THIS FINDING WAS INCONSISTENT WITH THE VERY IDEA OF AN ETHER AND SEEMED TO VIOLATE THE GALILEAN THEORY OF RELATIVITY.
HUH?!
SO WAS IT APPARENT THAT THE SPEED OF LIGHT WAS CONSTANT AND DID NOT CHANGE EVEN WHEN OBSERVED FROM A MOVING OBJECT?
THE SPEED OF LIGHT IS 300,000 KM/S WHEN OBSERVED FROM A STATE OF REST.
THE SPEED OF LIGHT IS 300,000 KM/S EVEN WHEN OBSERVED FROM A MOVING STATE.
MMMHMMM. THE CONSTANT SPEED OF LIGHT WAS A SERIOUS MATTER, AND IT COULD NOT BE EXPLAINED BY THE GALILEAN PRINCIPLE OF RELATIVITY.

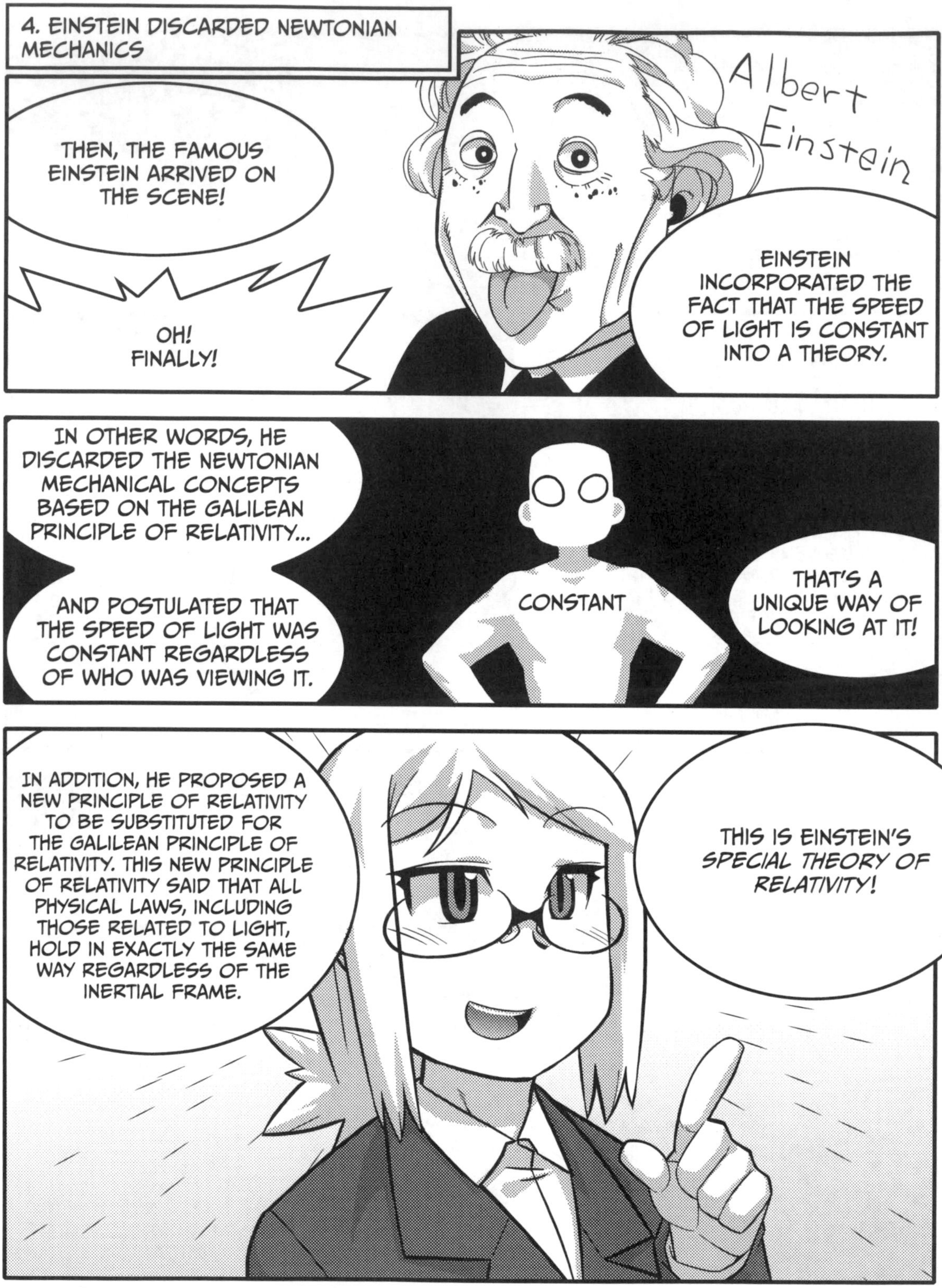
4. EINSTEIN DISCARDED NEWTONIAN MECHANICS
THEN, THE FAMOUS EINSTEIN ARRIVED ON THE SCENE!
OH! FINALLY!
Albert Einstein
EINSTEIN INCORPORATED THE FACT THAT THE SPEED OF LIGHT IS CONSTANT INTO A THEORY.
IN OTHER WORDS, HE DISCARDED THE NEWTONIAN MECHANICAL CONCEPTS BASED ON THE GALILEAN PRINCIPLE OF RELATIVITY...
AND POSTULATED THAT THE SPEED OF LIGHT WAS CONSTANT REGARDLESS OF WHO WAS VIEWING IT.
CONSTANT
THAT'S A UNIQUE WAY OF LOOKING AT IT!
IN ADDITION, HE PROPOSED A NEW PRINCIPLE OF RELATIVITY TO BE SUBSTITUTED FOR THE GALILEAN PRINCIPLE OF RELATIVITY. THIS NEW PRINCIPLE OF RELATIVITY SAID THAT ALL PHYSICAL LAWS, INCLUDING THOSE RELATED TO LIGHT, HOLD IN EXACTLY THE SAME WAY REGARDLESS OF THE INERTIAL FRAME.
THIS IS EINSTEIN'S *SPECIAL THEORY OF RELATIVITY!*

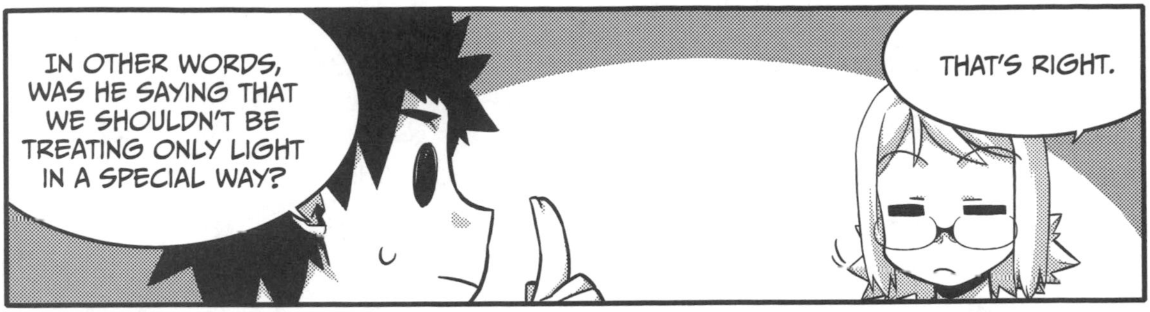
IN OTHER WORDS, WAS HE SAYING THAT WE SHOULDN'T BE TREATING ONLY LIGHT IN A SPECIAL WAY?
THAT'S RIGHT.

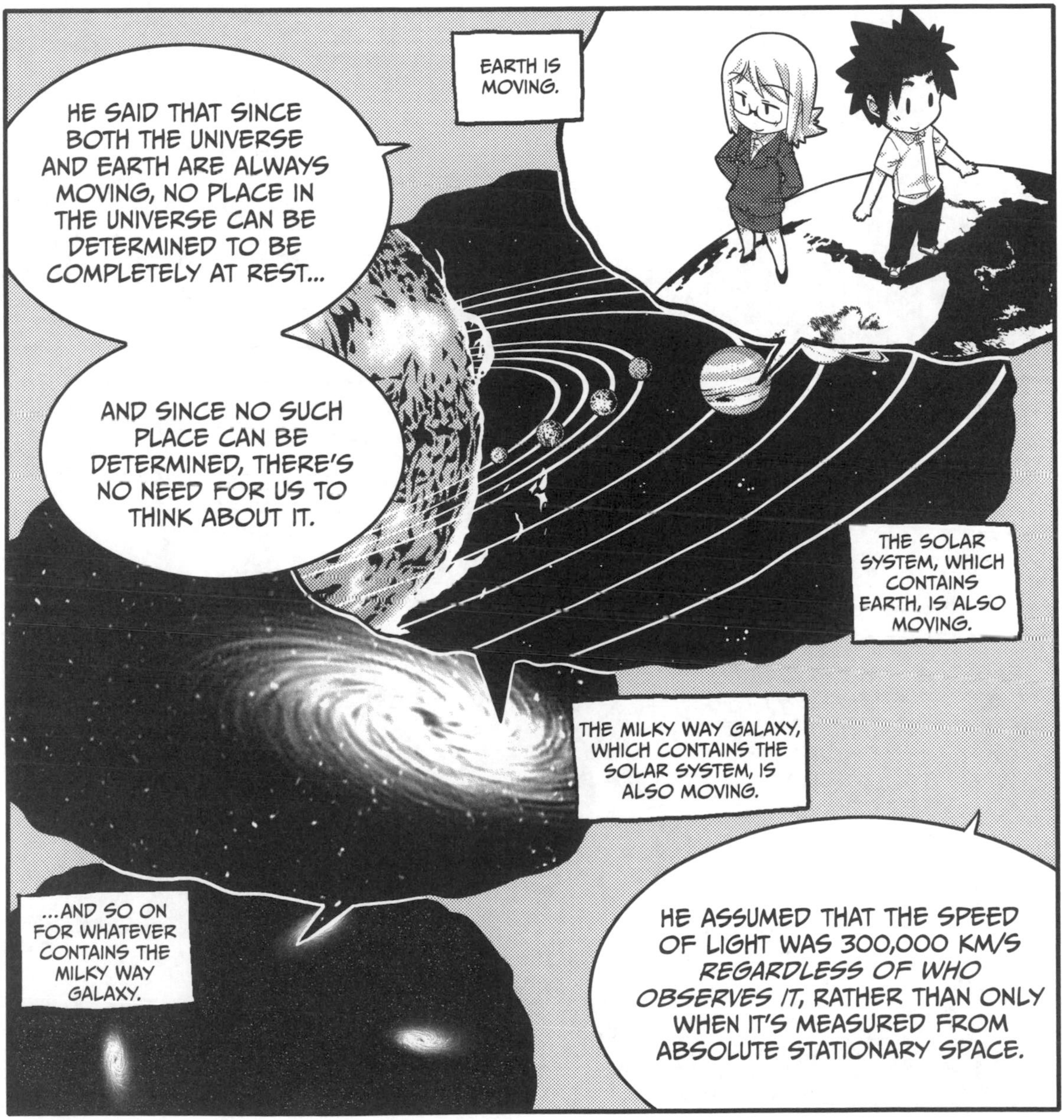
EARTH IS MOVING.
HE SAID THAT SINCE BOTH THE UNIVERSE AND EARTH ARE ALWAYS MOVING, NO PLACE IN THE UNIVERSE CAN BE DETERMINED TO BE COMPLETELY AT REST...
AND SINCE NO SUCH PLACE CAN BE DETERMINED, THERE'S NO NEED FOR US TO THINK ABOUT IT.
THE SOLAR SYSTEM, WHICH CONTAINS EARTH, IS ALSO MOVING.
THE MILKY WAY GALAXY, WHICH CONTAINS THE SOLAR SYSTEM, IS ALSO MOVING.
...AND SO ON FOR WHATEVER CONTAINS THE MILKY WAY GALAXY.
HE ASSUMED THAT THE SPEED OF LIGHT WAS 300,000 KM/S *REGARDLESS OF WHO OBSERVES IT*, RATHER THAN ONLY WHEN IT'S MEASURED FROM ABSOLUTE STATIONARY SPACE.

IN NEWTONIAN MECHANICS, LENGTH AND TIME HAD BEEN CONSIDERED TO BE ABSOLUTE CONCEPTS.

EVEN NOW, THEY FEEL THAT WAY IN EVERYDAY LIFE.

IF THEY WEREN'T, THEN WE COULDN'T DETERMINE APPOINTMENT TIMES OR MEASURE LENGTHS WITH RULERS, RIGHT?

NEVERTHELESS, EINSTEIN OVERTURNED THOSE CONCEPTS!

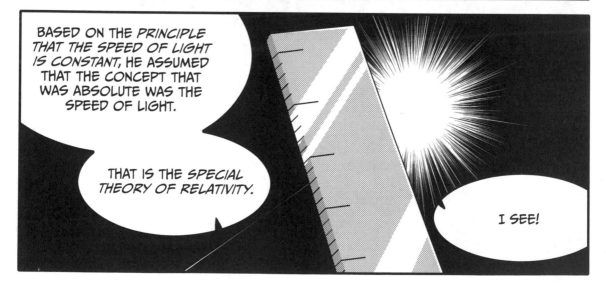

BASED ON THE *PRINCIPLE THAT THE SPEED OF LIGHT IS CONSTANT*, HE ASSUMED THAT THE CONCEPT THAT WAS ABSOLUTE WAS THE SPEED OF LIGHT.

THAT IS THE *SPECIAL THEORY OF RELATIVITY*.

I SEE!

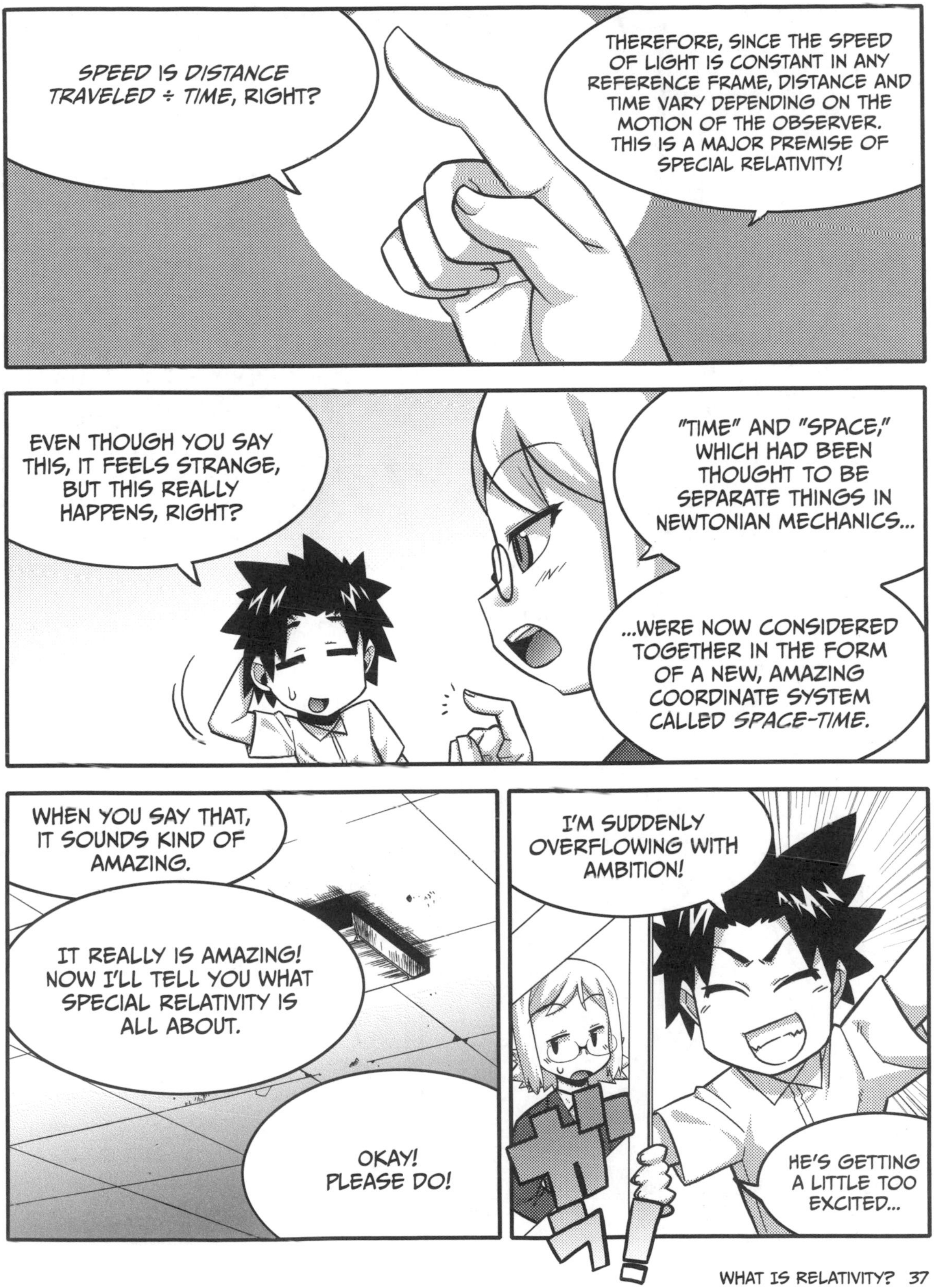
SPEED IS DISTANCE TRAVELED ÷ TIME, RIGHT?
THEREFORE, SINCE THE SPEED OF LIGHT IS CONSTANT IN ANY REFERENCE FRAME, DISTANCE AND TIME VARY DEPENDING ON THE MOTION OF THE OBSERVER. THIS IS A MAJOR PREMISE OF SPECIAL RELATIVITY!
EVEN THOUGH YOU SAY THIS, IT FEELS STRANGE, BUT THIS REALLY HAPPENS, RIGHT?
"TIME" AND "SPACE," WHICH HAD BEEN THOUGHT TO BE SEPARATE THINGS IN NEWTONIAN MECHANICS...
...WERE NOW CONSIDERED TOGETHER IN THE FORM OF A NEW, AMAZING COORDINATE SYSTEM CALLED SPACE-TIME.
WHEN YOU SAY THAT, IT SOUNDS KIND OF AMAZING.
IT REALLY IS AMAZING! NOW I'LL TELL YOU WHAT SPECIAL RELATIVITY IS ALL ABOUT.
OKAY! PLEASE DO!
I'M SUDDENLY OVERFLOWING WITH AMBITION!
HE'S GETTING A LITTLE TOO EXCITED...

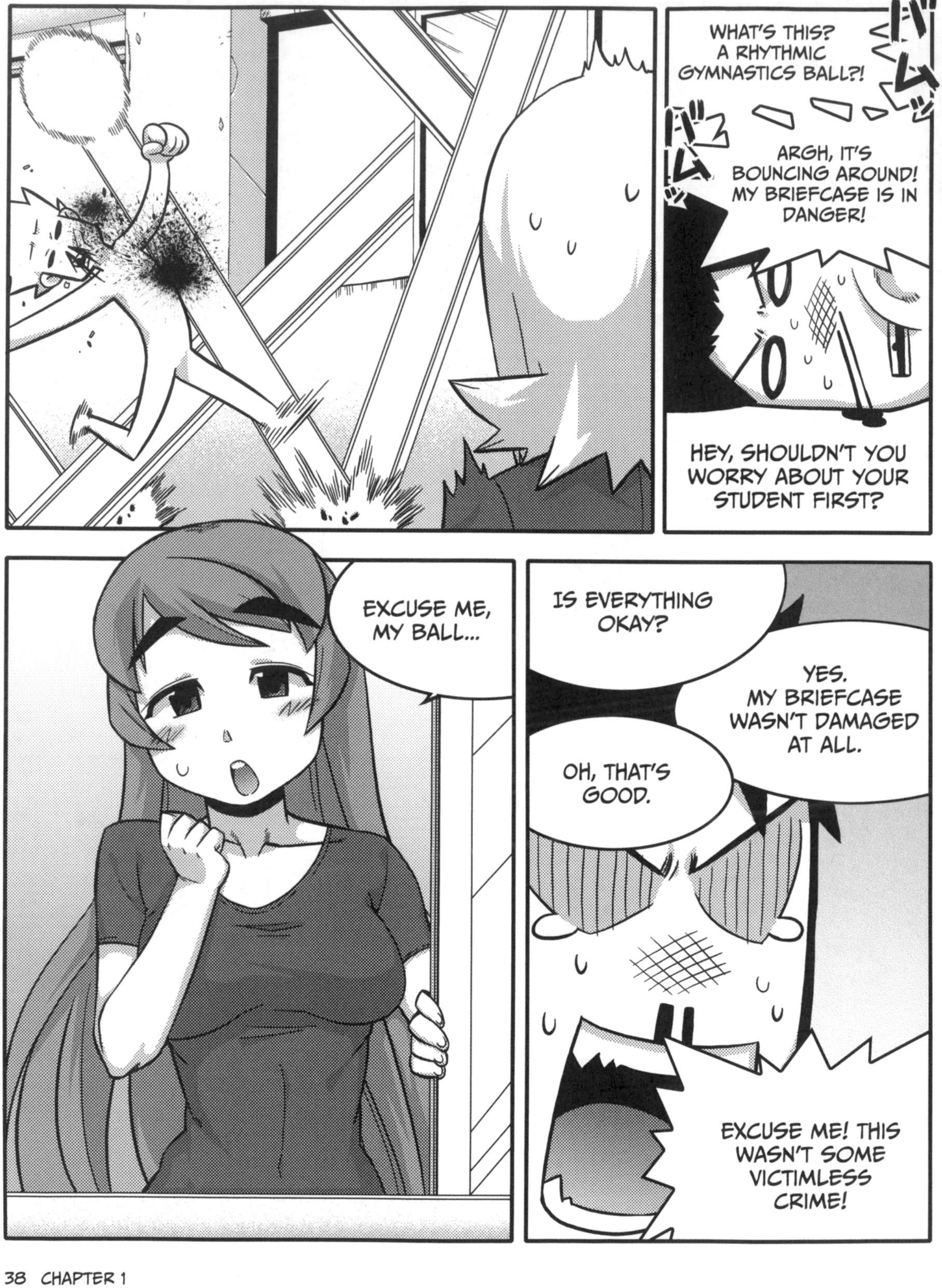
WHAT'S THIS? A RHYTHMIC GYMNASTICS BALL?!
ARGH, IT'S BOUNCING AROUND! MY BRIEFCASE IS IN DANGER!
HEY, SHOULDN'T YOU WORRY ABOUT YOUR STUDENT FIRST?
EXCUSE ME, MY BALL...
IS EVERYTHING OKAY?
YES. MY BRIEFCASE WASN'T DAMAGED AT ALL.
OH, THAT'S GOOD.
EXCUSE ME! THIS WASN'T SOME VICTIMLESS CRIME!

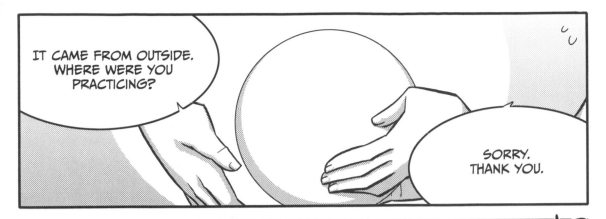

IT CAME FROM OUTSIDE. WHERE WERE YOU PRACTICING?

SORRY. THANK YOU.

MINAGI, YOU SHOULD ALSO STICK TO THE TOPIC, RIGHT? ♪

HUH? WHAT DO YOU KNOW...

...OMIGOSH, SHE KNOWS MY NAME!

YOU DID SPEAK UP IN FRONT OF THE ENTIRE STUDENT BODY.

...AH

WHAT IS LIGHT?

Maxwell's equations tell us that light is an electromagnetic wave. The color of light is determined by the wavelength of the electromagnetic wave. Red light has a wavelength of 630 nm, and blue light has a shorter wavelength of approximately 400 nm, where one nanometer (1 nm) = one billionth of a meter (10^{-9} m). Electromagnetic radiation at different wavelengths takes many forms, such as radio waves, X-rays, and gamma (γ) rays (see Figure 1-1).

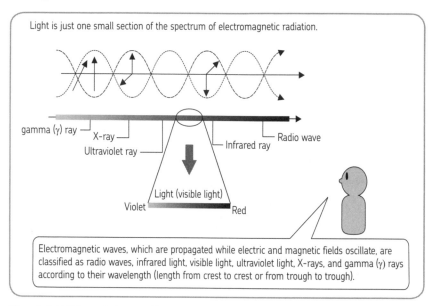

Light is just one small section of the spectrum of electromagnetic radiation.

gamma (γ) ray
X-ray
Ultraviolet ray
Infrared ray
Radio wave
Light (visible light)
Violet
Red

Electromagnetic waves, which are propagated while electric and magnetic fields oscillate, are classified as radio waves, infrared light, visible light, ultraviolet light, X-rays, and gamma (γ) rays according to their wavelength (length from crest to crest or from trough to trough).

Figure 1-1: Light is an electromagnetic wave.

Although light may seem common enough—it is all around us, after all—it is fundamental to both relativity and quantum theory, the cornerstones of modern physics.

But before we delve into light's true nature, let's introduce the properties of light that have been known for a long time.

First, you know that light is *reflected* by a mirror or the surface of water. You also know about the *refraction* of light—you only need to look at your feet the next time you take a bath or see how your straw "bends" when you put it in a glass of water. Any change in medium changes a wave's direction, due to a change in the wave's speed through that medium.

Some mediums refract light of different wavelengths different amounts. In other words, light of different colors is bent to different degrees, a property known as *dispersion*. This causes white light, which consists of light of all colors, to be spread out into a spectrum of light from red to violet. We can see the seven colors of a rainbow because of dispersion.

These properties of reflection, refraction, and dispersion have been used to create precision camera lenses and telescopes. Figure 1-2 shows what happens to light when it is reflected, refracted, or dispersed.

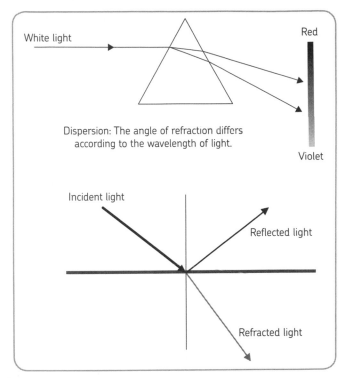

White light
Red
Violet
Dispersion: The angle of refraction differs according to the wavelength of light.
Incident light
Reflected light
Refracted light

Figure 1-2: Dispersion, reflection, and refraction

Next, more subtle phenomena called *interference* and *diffraction* can be observed. These phenomena stem from the fact that light is a wave. Interference describes what happens when two light waves come together. When the two waves come together, the result is either *constructive interference*, where the waves' amplitudes are added together, or *destructive interference*, where one wave's amplitude is subtracted from the other's. Figure 1-3 shows the different kinds of interference.

Interference between light waves
+
=
If a crest is superimposed upon a crest, there will be a larger crest.
+
=
If a crest is superimposed upon a trough, they will cancel each other out and become flat.

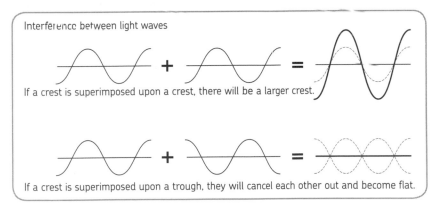

Figure 1-3: Interference can make waves stronger or weaker.

Diffraction can be observed when light passes through a tiny hole about the same size as the wavelength of the light. Due to the constructive and destructive interference of different parts of the light wave with itself, passing through a tiny aperture can cause the light to spread out or bend, as shown in Figure 1-4. Diffraction is often what limits the resolution of microscopes.

Diffraction: Light passes through a small hole (aperture) and bends and spreads out.
Light
Small hole
A diffraction pattern appears

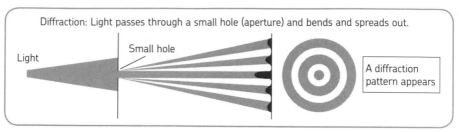

Figure 1-4: Diffraction comes about from interference.

Another property of light is called *polarization,* a property that describes the orientation of the transverse electric and magnetic components of the electromagnetic wave. This property is very useful; it allows special filters to be made (called *polarizing filters*) that allow only light with a specific polarization to pass (see Figure 1-5).

Polarizing filter
Only the vertical component emerges.

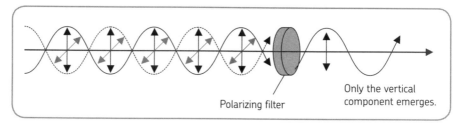

Figure 1-5: Polarization

In *scattering,* light collides with dust and other particles in the air, thereby changing direction (see Figure 1-6). Since blue light (with shorter wavelengths) is scattered by water molecules in the air more than red light (with longer wavelengths), the sky appears blue.

Incident light
Scattering light

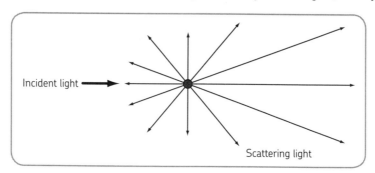

Figure 1-6: Scattering

LIGHT IS CONSTANT (AND THEY PROVE IT EVERY DAY IN A LAB CALLED SPRING-8)

Various tests have been conducted to verify that the speed of light is truly constant. This is important because it is one of the fundamental premises of relativity.

One way that we can test this property is to measure the speed of light coming from an object that is moving very fast. If the speed of light is not constant, the Newtonian notion of "adding" relative velocities predicts that light coming from an object moving towards the observer will be the speed of light plus the speed of the moving object; for example, if the object is moving near the speed of light, then the light from the object should be moving nearly twice the speed of light. If the speed of light is constant, on the other hand, than the light coming from the fast-moving object will just be the speed of light. Measurements confirm that the speed of light is always the same, regardless of the speed of the object from which it comes (see Figure 1-7).

Light emitted from electrons that are traveling at nearly the speed of light is observed to be moving at the speed of light.
Electron traveling at nearly the speed of light emits light.
The light emitted by the electron is at the speed of light rather than the "speed of the electron + speed of light."

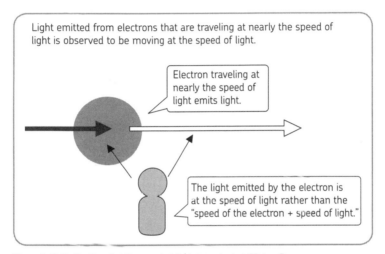

Figure 1-7: Verification that the speed of light is constant at SPring-8

Moving objects near the speed of light for these experiments is extremely difficult, and these experiments are performed at very specialized scientific facilities. SPring-8 is a synchotron radiation facility in Japan's Hyogo Prefecture that performs experiments by smashing together electrons traveling at extremely fast speeds (99.9999998 percent of the speed of light). Besides verifying that the speed of light is constant, these experiments help scientists uncover the basic building blocks of matter.

WHAT'S SIMULTANEOUS DEPENDS ON WHOM YOU ASK!
(SIMULTANEITY MISMATCH)

If we consider the principle that "the speed of light is constant," various phenomena appear strange. One of these is the phenomenon called the *simultaneity mismatch*, which means that what is simultaneous for me is not the same as what is simultaneous for you.

I can imagine that you are thinking, "What in the world are you saying?" So let's consider the concept of "simultaneous" again. We will compare the case of Newtonian velocity addition (nonrelativistic addition of velocity) with the case in which the speed of light is constant (relativistic addition of velocity).

Consider Mr. A, who is riding on a rocket flying at a constant velocity, and Mr. B, who is observing Mr. A from a stationary space station. Assume that Mr. A is in the middle of the rocket. Sensors have been placed at the front and back of the rocket. Mr. A throws balls (or emits light) toward the front and back of the rocket. We will observe how those balls (or light beams) hit the sensors at the front and back of the rocket.

CASE OF NEWTONIAN VELOCITY ADDITION
(NONRELATIVISTIC ADDITION)

First, we will use the motion of the balls to consider the case in which velocities are added in a Newtonian mechanical manner (before considering relativity).

First, let's look at Mr. A as shown in Figure 1-8. Since from Mr. A's perspective the rocket is not moving, the balls, which are moving at the same velocity from the center toward the sensors at the front and back of the rocket, arrive at the sensors "simultaneously."

Next, when observed by Mr. B from the space station, the rocket advances in the direction of travel. In other words, using the point of departure of the balls (dotted line) as a reference, the front of the ship moves away from the dotted line, and the back of the ship approaches the dotted line. However, since the velocity of the rocket is added to the velocity of the ball in the forward direction, according to normal addition, the ball's velocity increases and it catches up with the front of the ship. On the other hand, the velocity of the ball toward the back of the ship is reduced by the velocity of the rocket (indicated by the short arrow in the figure), and the back of the ship catches up to the ball. Therefore, Mr. B also observes that the balls arrive at the front and back of the ship "simultaneously."

Nonrelativistic addition:

Mr. A observes the motion of the balls inside the rocket.

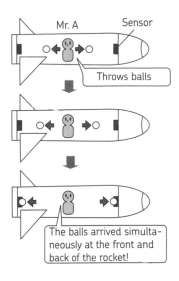

Mr. A
Sensor
Throws balls
The balls arrived simulta-
neously at the front and
back of the rocket!

Nonrelativistic addition:

Mr. B observes the motion of the balls inside the rocket from his space station. Since the balls are moving together with the rocket, the velocity of the ball is increased by the velocity of the rocket toward the front of the rocket and decreased by the velocity of the rocket toward the back of the rocket. Therefore, the balls arrive "simultaneously" (the lengths of the arrows indicate the difference in the velocities of the balls).

Rocket velocity
Mr. B
The balls arrived "simultaneously" at the front and
back of the rocket!

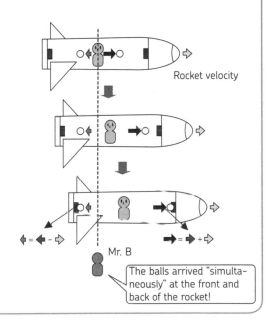

Figure 1-8: Newtonian velocity addition

CASE IN WHICH THE SPEED OF LIGHT IS CONSTANT (RELATIVISTIC ADDITION OF VELOCITY)

Now let's consider the case in which the speed of light is constant. Instead of throwing balls, Mr. A will emit light while traveling at nearly the speed of light (see Figure 1-9).

When the speed of light is constant: Mr. A observes the motion of the light inside the rocket.
When the speed of light is constant: Mr. B observes the motion of the light from the space station. Since the light is moving at a constant speed, it will arrive first at the back of the ship and not arrive at the front of the ship for a long time.
Mr. A
Sensor
Emits light
The light arrived simultaneously at the front and back of the rocket!
Rocket velocity
Mr. B on space station
The light arrived at the back of the rocket first! Since it still hasn't arrived at the front of the rocket, its arrival was not "simultaneous"!

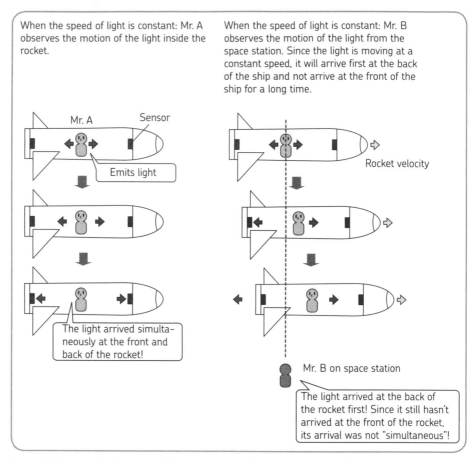

Figure 1-9: Case in which the speed of light is constant (relativistic addition of velocity)

You may have already realized what is at issue: Mr. B's observation will differ from that of Mr. A.

For Mr. A, even when the speed of light is constant, the light will arrive "simultaneously" at the front and back of the rocket.

However, when observed by Mr. B, the light moving towards the front of the ship does not arrive for a long time. It has to overtake the ship, which is moving away at nearly the speed of the light. Therefore, the light arrives at the back of the ship before it reaches the front of the ship.

That's right; when observed by Mr. B, the light does not arrive "simultaneously" at the front and back of the ship.

The simultaneity property of light differs in this way depending on the standpoint of the observer. This is called *simultaneity mismatch*.

GALILEAN PRINCIPLE OF RELATIVITY AND GALILEAN TRANSFORMATION

The Galilean principle of relativity says that "the laws of physics are the same regardless of whether the coordinate system from which the observation is made is at rest or moving at a constant velocity." In other words, Newtonian mechanics (the physical laws that govern motion) are always the same, regardless of whether observations are made in a reference frame that is at rest or one that is moving at a constant velocity. This principle was derived from an experiment in which an iron ball was dropped from the mast of a ship, as shown in Figure 1-10. The iron ball fell directly under the mast whether the ship was moving or at rest.

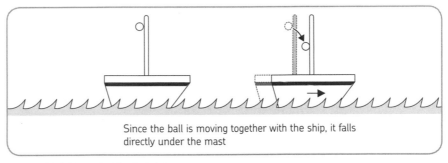

Since the ball is moving together with the ship, it falls directly under the mast

Figure 1-10: Galilean principle of relativity

Since the laws of physics are the same in any reference frame, Galileo arrived at a straightforward way to describe how observations look different depending on which reference frame you are in. Today we use algebraic equations called the *Galilean transformation* to help understand the notion of "adding" relative velocities.

Let's take two coordinate systems, one with the coordinates (x, t) and the other with coordinates (x', t'), where x and x' describe position and t and t' describe time. One can go from one coordinate system to the other, by considering the relative velocity between the two coordinate systems v.

$$x' = x - vt$$
$$t' = t$$

The above equations show the relationship between coordinates from a coordinate system at rest and a coordinate system moving at a constant velocity v relative to the coordinate system at rest. Inertial frames are mutually linked in this way by the Galilean transformation. If we compare them using Newton's equation of motion, we can prove that Newton's equation of motion takes the same form in each inertial frame. In other words, when the Galilean principle of relativity holds, Newtonian mechanics will hold.

DIFFERENCES BETWEEN THE GALILEAN PRINCIPLE OF RELATIVITY AND EINSTEIN'S SPECIAL PRINCIPLE OF RELATIVITY

As just described, the Galilean principle of relativity indicates that Newtonian mechanics apply across inertial frames when linked with the Galilean transformation.

On the other hand, the assumption that the speed of light is constant in any reference frame forced scientists to reformulate the Galilean transformation to be consistent with relativity. This new transformation is called the *Lorentz transformation*.

The Lorentz transformation is shown by the equations below, which show the relationship between coordinates from a coordinate system at rest and a coordinate system moving at a constant velocity v relative to the coordinate system at rest. The variables with the prime symbol (') attached represent coordinates observed from the coordinate system at rest; the variables without the prime symbol represent coordinates observed from the system in motion. Note that the speed of light c appears in the equations here. Another point to notice is that time t is transformed in a manner similar to that of length; time does not exist independently but must be considered to be unified with space.

$$x' = \frac{x - vt}{\sqrt{1 - \left(\frac{v}{c}\right)^2}}$$

$$t' = \frac{t - \frac{v}{c^2}x}{\sqrt{1 - \left(\frac{v}{c}\right)^2}}$$

WAIT A SECOND—WHAT HAPPENS WITH THE ADDITION OF VELOCITIES?

When we assume that the speed of light is constant, what happens when velocities are added to the mix?

According to the principle of relativity, when calculated based on the Lorentz transformation, the addition of velocities is indicated by the following equation.

$$w = \frac{u + v}{1 + \frac{vu}{c^2}}$$

This equation describes the resulting addition of velocities of a missile w when the velocity of a rocket is v and the velocity (observed from the rocket) of the missile shot from the rocket is u, as shown in Figure 1-11. The difference is apparent when this equation is compared with the normal addition (nonrelativistic) equation $w = u + v$.

If we enter specific velocities in the above equations, we'll obtain some interesting results.

In the nonrelativistic case, when Mr. B on the space station observes the missile that was fired from the rocket, if v denotes the velocity of the rocket and u denotes the velocity of the missile observed from the rocket, then the addition of velocities is indicated by w = u + v.
The missile has velocity u.
Mr. A
Rocket velocity v
With the principle of relativity, the addition of velocities is given by
$w = \frac{u+v}{1+\frac{vu}{c^2}}$
Mr. A's rocket has velocity v.
Mr. B on space station

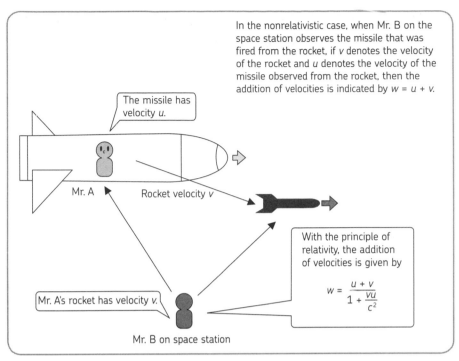

Figure 1-11: Addition of velocities

For example, when the rocket velocity v is 50 percent of the speed of light ($0.5c$) and the missile velocity u observed from the rocket is 50 percent of the speed of light (also $0.5c$), then the missile velocity w observed by Mr. B will be 80 percent of the speed of light ($0.8c$).

$$w = \frac{(0.5c + 0.5c)}{\left(1 + \frac{(0.5c)^2}{c^2}\right)} = \frac{c}{1.25} = 0.8c$$

This equation also yields an interesting result when v and u are their maximum values. If the rocket velocity v is 100 percent of the speed of light (practically speaking, $v = c$ is impossible for an object with mass, like a rocket) and the missile velocity u observed from the rocket is 100 percent of the speed of light, then the missile velocity w observed by Mr. B will be the speed of light.

$$w = \frac{(c + c)}{\left(1 + \frac{c^2}{c^2}\right)} = \frac{2c}{2} = c$$

The speed of light cannot be exceeded under any circumstances!

WHAT DO YOU MEAN, TIME SLOWS DOWN?

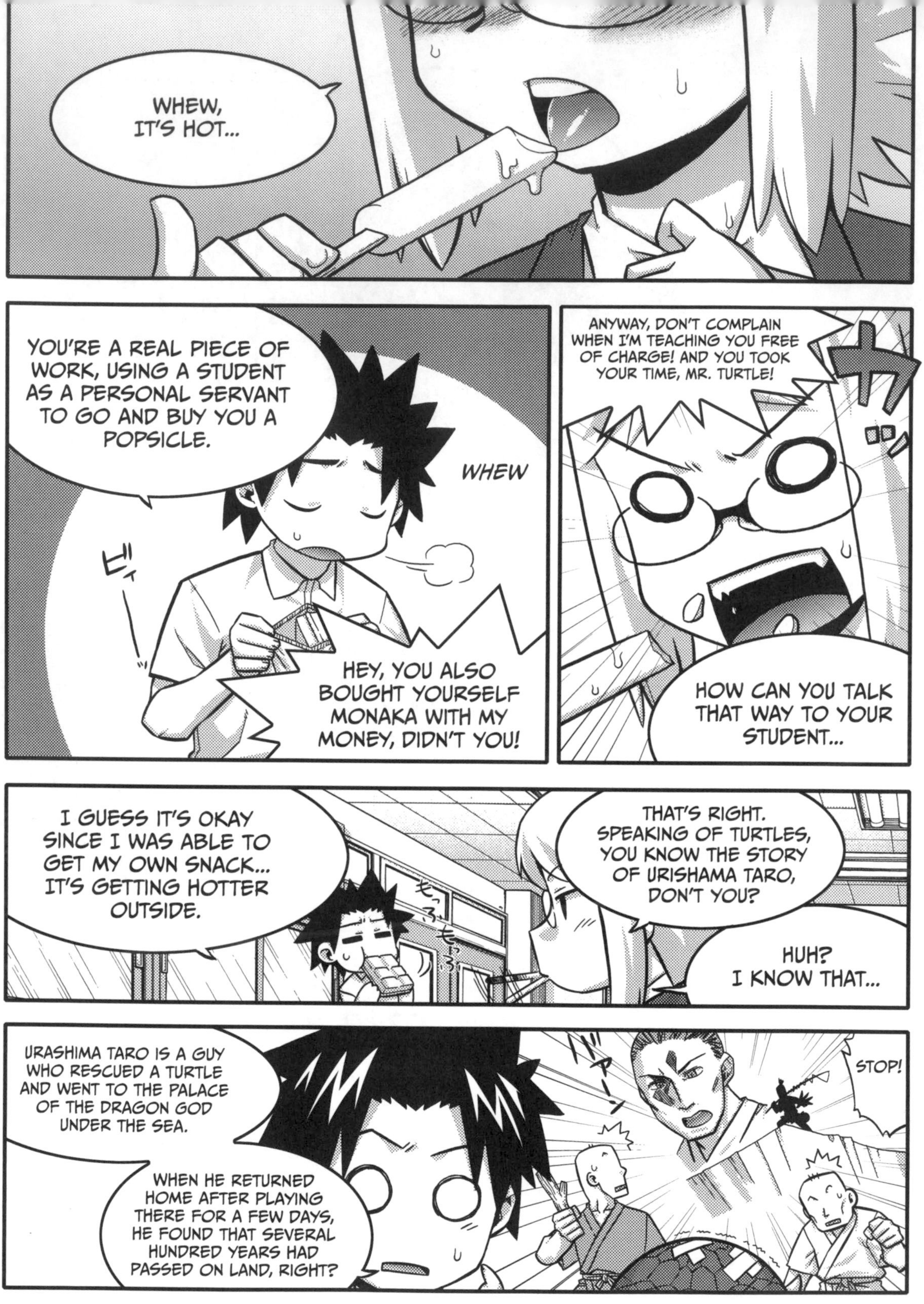

WHEW, IT'S HOT...
YOU'RE A REAL PIECE OF WORK, USING A STUDENT AS A PERSONAL SERVANT TO GO AND BUY YOU A POPSICLE.
WHEW
HEY, YOU ALSO BOUGHT YOURSELF MONAKA WITH MY MONEY, DIDN'T YOU!
ANYWAY, DON'T COMPLAIN WHEN I'M TEACHING YOU FREE OF CHARGE! AND YOU TOOK YOUR TIME, MR. TURTLE!
HOW CAN YOU TALK THAT WAY TO YOUR STUDENT...
I GUESS IT'S OKAY SINCE I WAS ABLE TO GET MY OWN SNACK... IT'S GETTING HOTTER OUTSIDE.
THAT'S RIGHT. SPEAKING OF TURTLES, YOU KNOW THE STORY OF URISHAMA TARO, DON'T YOU?
HUH? I KNOW THAT...
URASHIMA TARO IS A GUY WHO RESCUED A TURTLE AND WENT TO THE PALACE OF THE DRAGON GOD UNDER THE SEA.
WHEN HE RETURNED HOME AFTER PLAYING THERE FOR A FEW DAYS, HE FOUND THAT SEVERAL HUNDRED YEARS HAD PASSED ON LAND, RIGHT?
STOP!

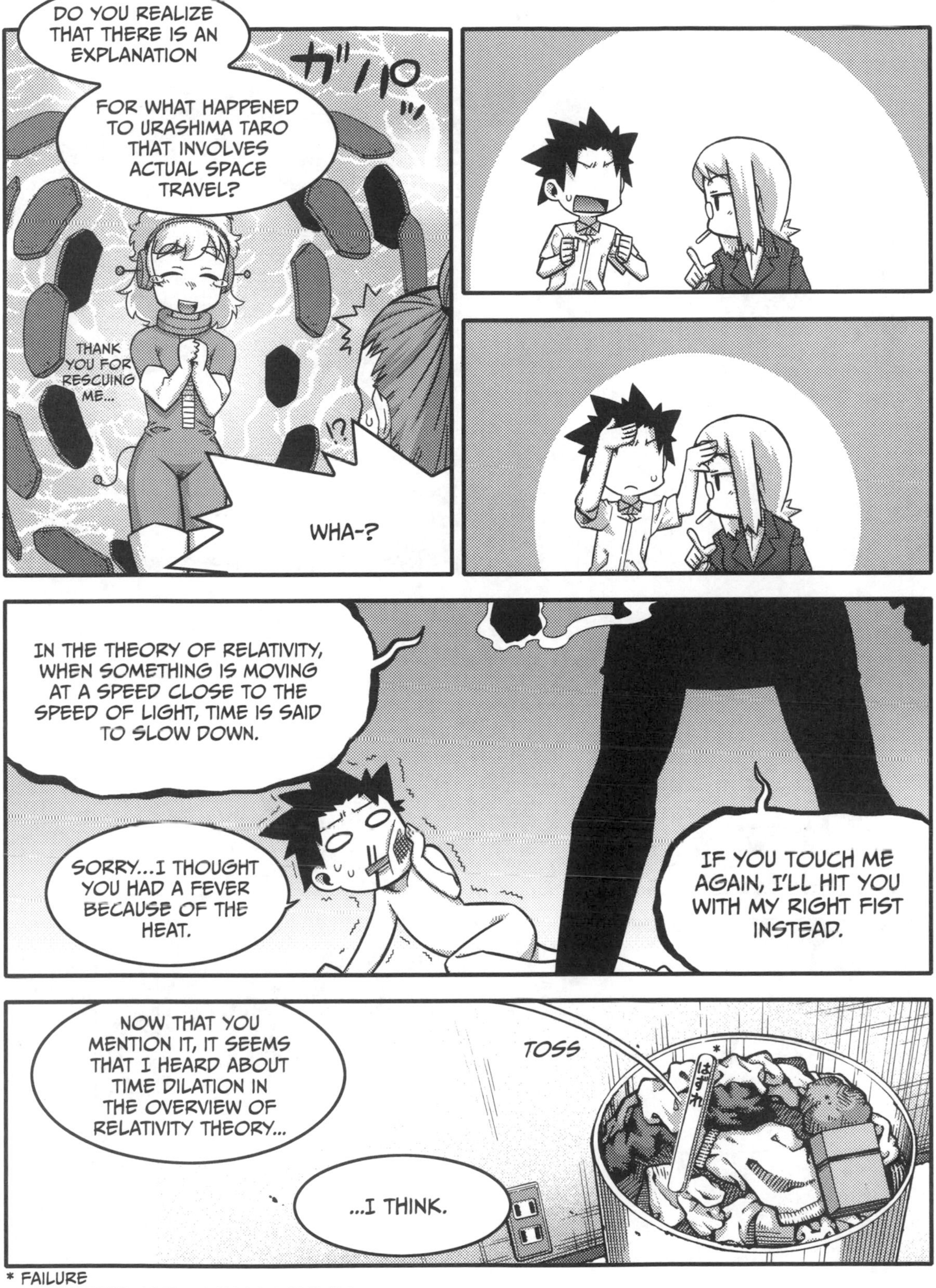
DO YOU REALIZE THAT THERE IS AN EXPLANATION
FOR WHAT HAPPENED TO URASHIMA TARO THAT INVOLVES ACTUAL SPACE TRAVEL?
THANK YOU FOR RESCUING ME...
!?
WHA-?
IN THE THEORY OF RELATIVITY, WHEN SOMETHING IS MOVING AT A SPEED CLOSE TO THE SPEED OF LIGHT, TIME IS SAID TO SLOW DOWN.
SORRY...I THOUGHT YOU HAD A FEVER BECAUSE OF THE HEAT.
IF YOU TOUCH ME AGAIN, I'LL HIT YOU WITH MY RIGHT FIST INSTEAD.
NOW THAT YOU MENTION IT, IT SEEMS THAT I HEARD ABOUT TIME DILATION IN THE OVERVIEW OF RELATIVITY THEORY...
TOSS
* はずれ
...I THINK.

* FAILURE
TRANSLATOR'S NOTE: IN JAPAN, POPSICLE STICKS HAVE FORTUNES ON THEM.

1. URASHIMA EFFECT (TIME DILATION)

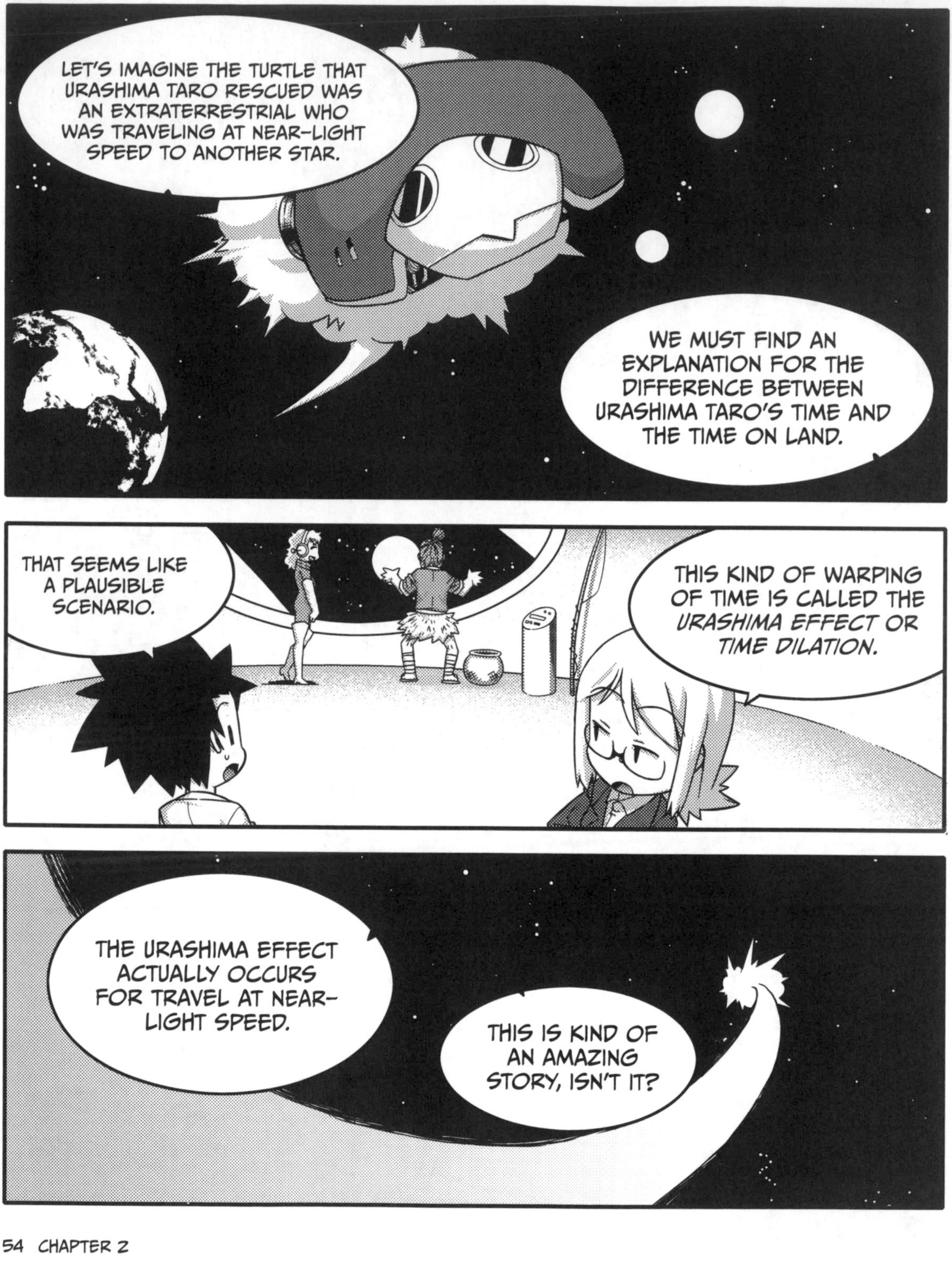
LET'S IMAGINE THE TURTLE THAT URASHIMA TARO RESCUED WAS AN EXTRATERRESTRIAL WHO WAS TRAVELING AT NEAR-LIGHT SPEED TO ANOTHER STAR.
WE MUST FIND AN EXPLANATION FOR THE DIFFERENCE BETWEEN URASHIMA TARO'S TIME AND THE TIME ON LAND.
THAT SEEMS LIKE A PLAUSIBLE SCENARIO.
THIS KIND OF WARPING OF TIME IS CALLED THE *URASHIMA EFFECT* OR *TIME DILATION*.
THE URASHIMA EFFECT ACTUALLY OCCURS FOR TRAVEL AT NEAR-LIGHT SPEED.
THIS IS KIND OF AN AMAZING STORY, ISN'T IT?

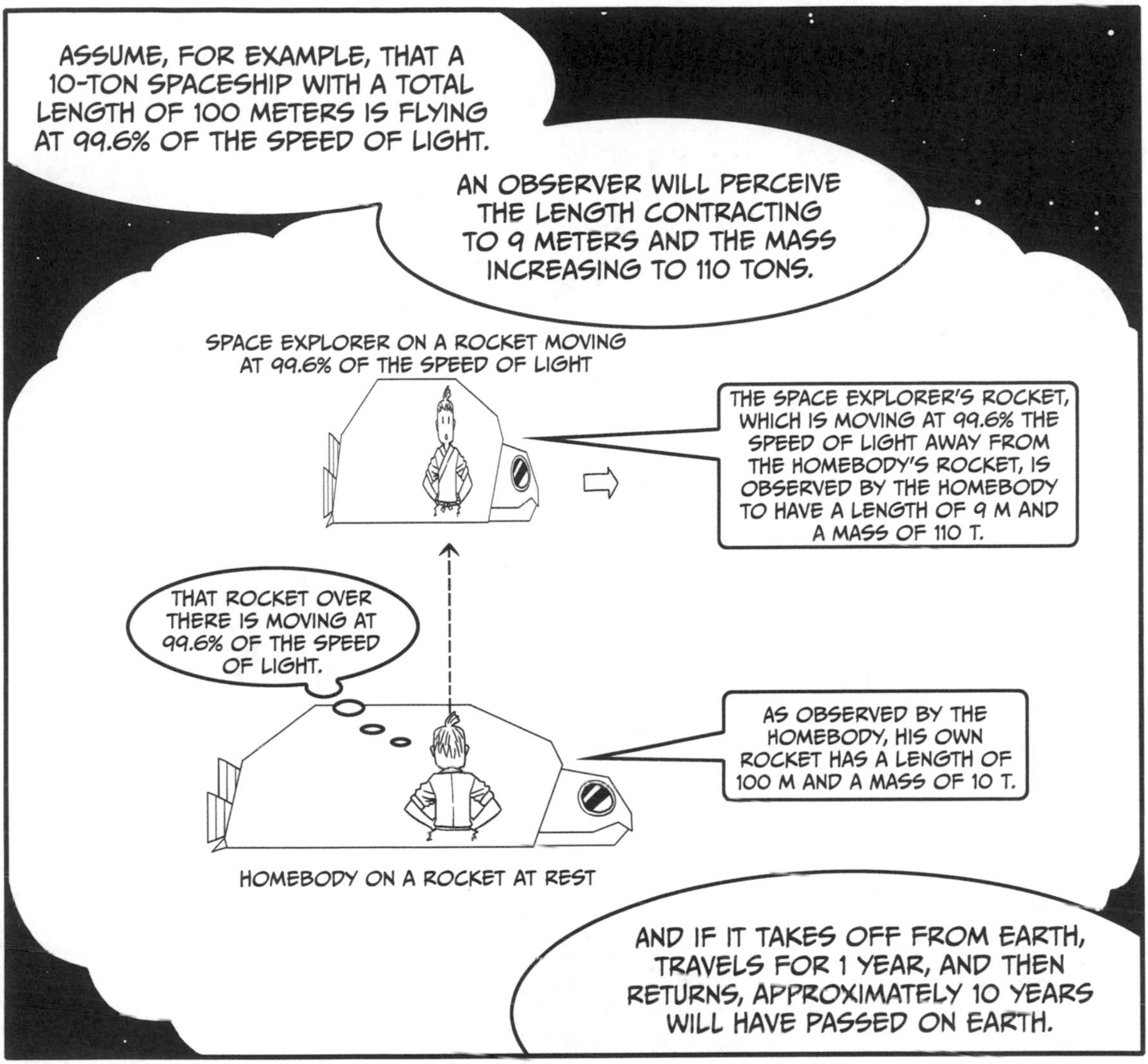
ASSUME, FOR EXAMPLE, THAT A 10-TON SPACESHIP WITH A TOTAL LENGTH OF 100 METERS IS FLYING AT 99.6% OF THE SPEED OF LIGHT.
AN OBSERVER WILL PERCEIVE THE LENGTH CONTRACTING TO 9 METERS AND THE MASS INCREASING TO 110 TONS.
SPACE EXPLORER ON A ROCKET MOVING AT 99.6% OF THE SPEED OF LIGHT
THE SPACE EXPLORER'S ROCKET, WHICH IS MOVING AT 99.6% THE SPEED OF LIGHT AWAY FROM THE HOMEBODY'S ROCKET, IS OBSERVED BY THE HOMEBODY TO HAVE A LENGTH OF 9 M AND A MASS OF 110 T.
THAT ROCKET OVER THERE IS MOVING AT 99.6% OF THE SPEED OF LIGHT.
AS OBSERVED BY THE HOMEBODY, HIS OWN ROCKET HAS A LENGTH OF 100 M AND A MASS OF 10 T.
HOMEBODY ON A ROCKET AT REST
AND IF IT TAKES OFF FROM EARTH, TRAVELS FOR 1 YEAR, AND THEN RETURNS, APPROXIMATELY 10 YEARS WILL HAVE PASSED ON EARTH.

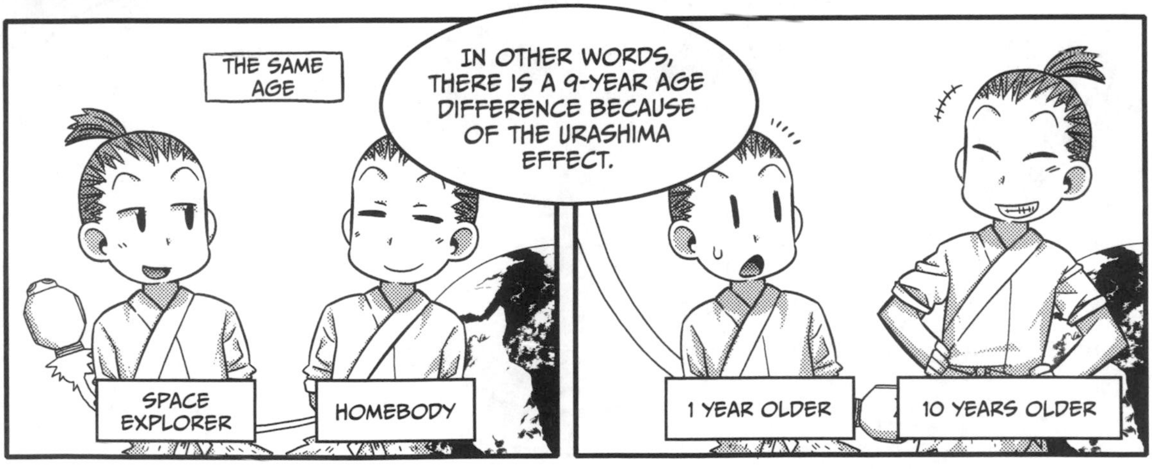
THE SAME AGE
IN OTHER WORDS, THERE IS A 9-YEAR AGE DIFFERENCE BECAUSE OF THE URASHIMA EFFECT.
SPACE EXPLORER
HOMEBODY
1 YEAR OLDER
10 YEARS OLDER

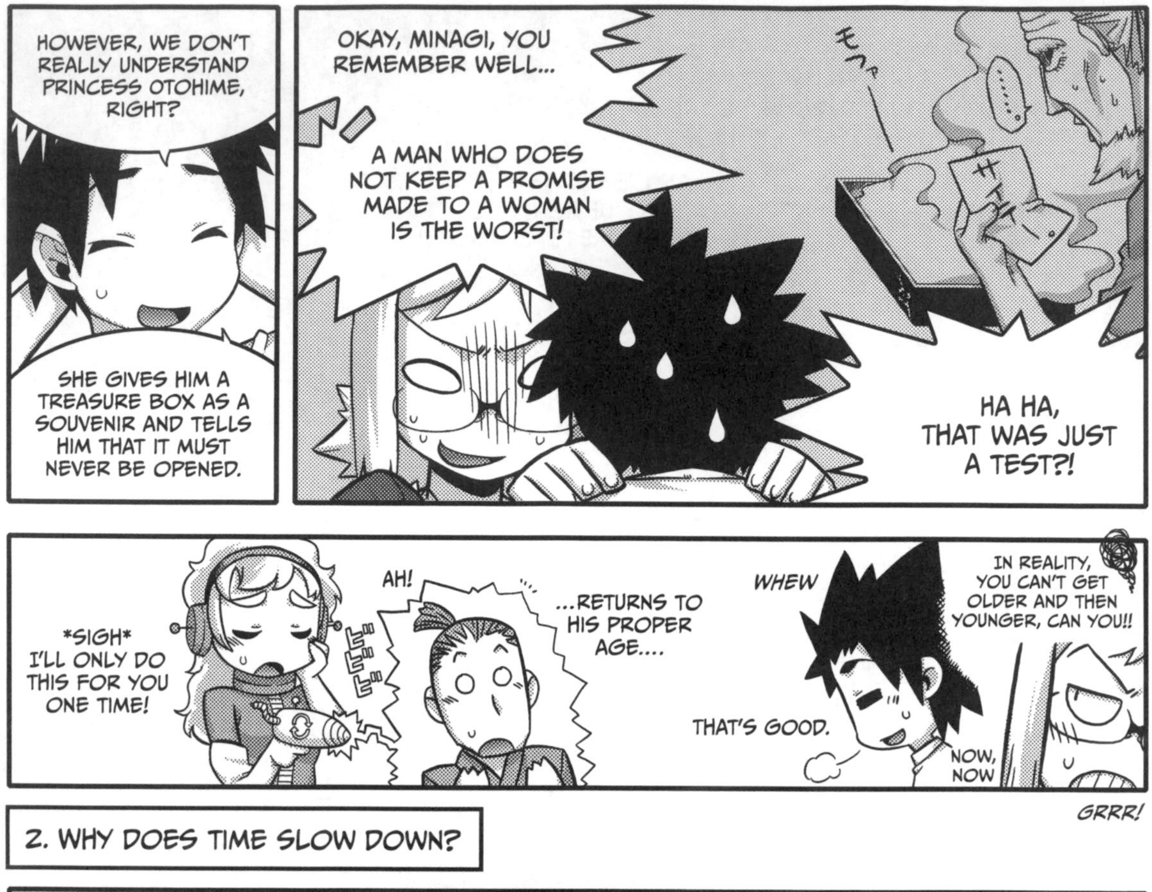
HOWEVER, WE DON'T REALLY UNDERSTAND PRINCESS OTOHIME, RIGHT?
SHE GIVES HIM A TREASURE BOX AS A SOUVENIR AND TELLS HIM THAT IT MUST NEVER BE OPENED.
OKAY, MINAGI, YOU REMEMBER WELL...
A MAN WHO DOES NOT KEEP A PROMISE MADE TO A WOMAN IS THE WORST!
HA HA, THAT WAS JUST A TEST?!
SIGH I'LL ONLY DO THIS FOR YOU ONE TIME!
AH!
...RETURNS TO HIS PROPER AGE....
WHEW
THAT'S GOOD.
IN REALITY, YOU CAN'T GET OLDER AND THEN YOUNGER, CAN YOU!!
NOW, NOW
GRRR!

2. WHY DOES TIME SLOW DOWN?

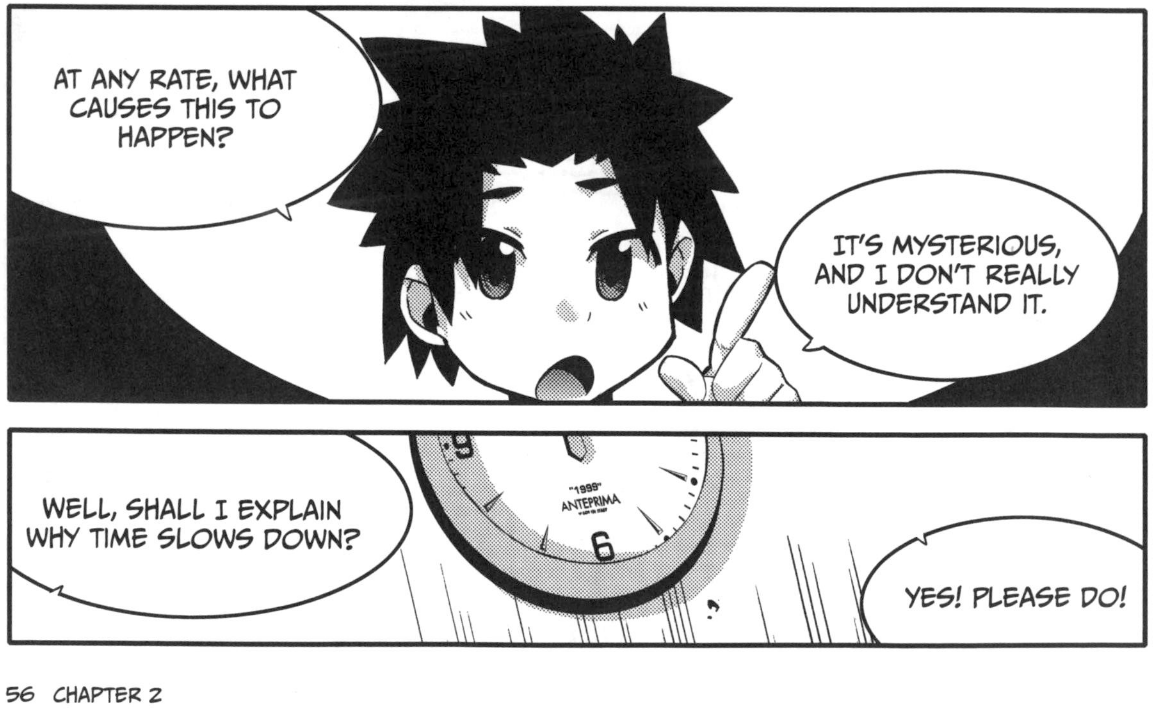
AT ANY RATE, WHAT CAUSES THIS TO HAPPEN?
IT'S MYSTERIOUS, AND I DON'T REALLY UNDERSTAND IT.
WELL, SHALL I EXPLAIN WHY TIME SLOWS DOWN?
YES! PLEASE DO!

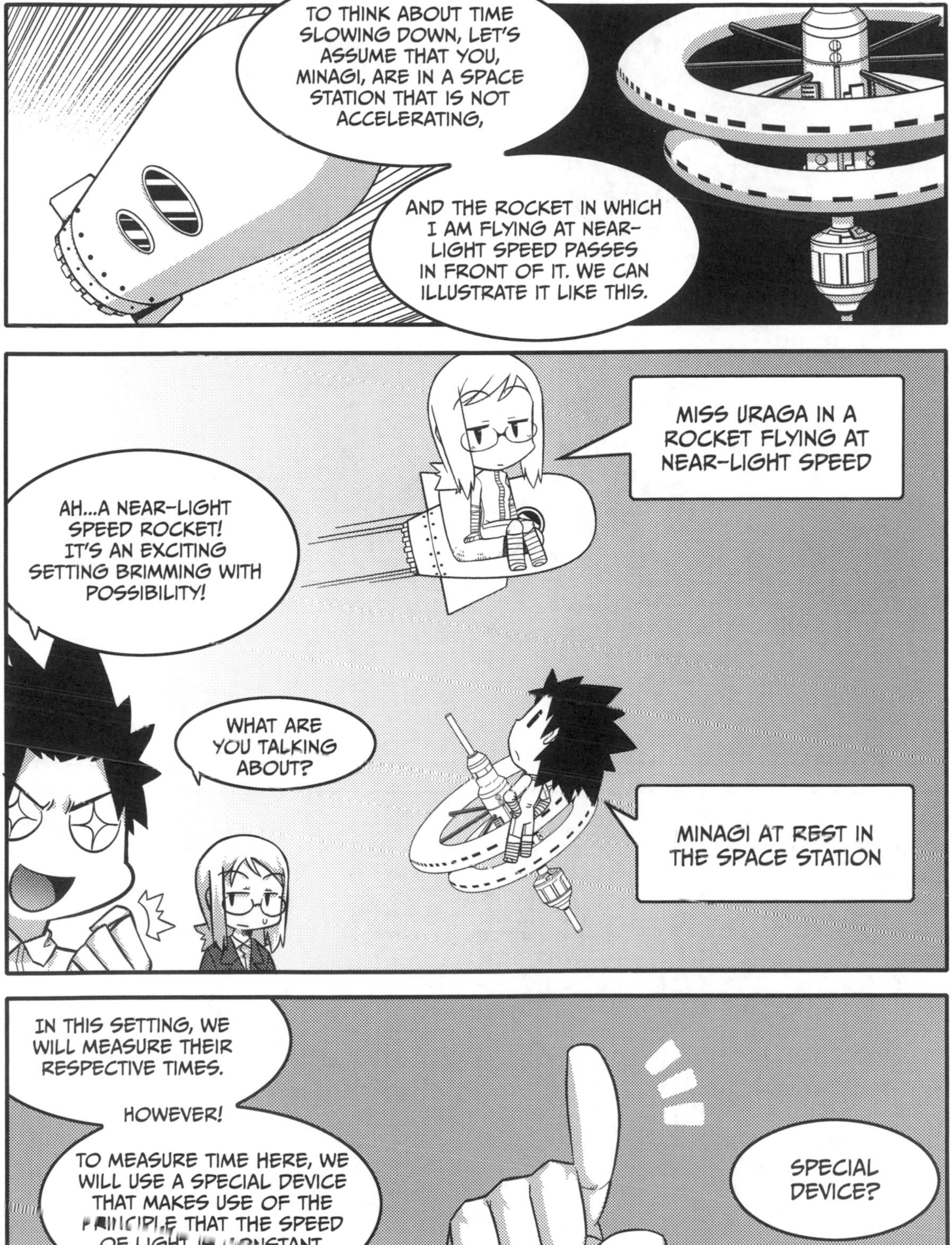
TO THINK ABOUT TIME SLOWING DOWN, LET'S ASSUME THAT YOU, MINAGI, ARE IN A SPACE STATION THAT IS NOT ACCELERATING,
AND THE ROCKET IN WHICH I AM FLYING AT NEAR-LIGHT SPEED PASSES IN FRONT OF IT. WE CAN ILLUSTRATE IT LIKE THIS.
AH...A NEAR-LIGHT SPEED ROCKET! IT'S AN EXCITING SETTING BRIMMING WITH POSSIBILITY!
MISS URAGA IN A ROCKET FLYING AT NEAR-LIGHT SPEED
WHAT ARE YOU TALKING ABOUT?
MINAGI AT REST IN THE SPACE STATION
IN THIS SETTING, WE WILL MEASURE THEIR RESPECTIVE TIMES.
HOWEVER!
TO MEASURE TIME HERE, WE WILL USE A SPECIAL DEVICE THAT MAKES USE OF THE PRINCIPLE THAT THE SPEED OF LIGHT IS CONSTANT.
SPECIAL DEVICE?

PLONK

IS THIS THE DEVICE YOU WERE TALKING ABOUT?

IT'S CALLED A LIGHT CLOCK!

IT'S A CYLINDRICAL DEVICE 30 CENTIMETERS LONG FOR MEASURING TIME USING LIGHT...

REALLY, IT'S JUST AN IMAGINARY DEVICE THAT IS USED TO FACILITATE THE EXPLANATION!

SOMEHOW THIS THING DOESN'T LOOK LIKE A CLOCK.

YOU'RE RIGHT, BUT IT'S REALLY IMPORTANT AND INDISPENSIBLE FOR OUR SUBSEQUENT EXPERIMENTS.

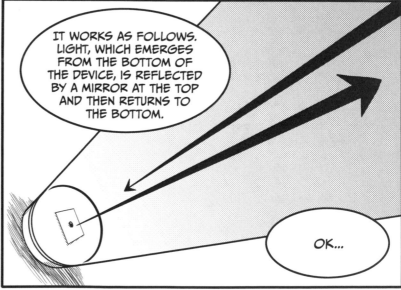

IT WORKS AS FOLLOWS. LIGHT, WHICH EMERGES FROM THE BOTTOM OF THE DEVICE, IS REFLECTED BY A MIRROR AT THE TOP AND THEN RETURNS TO THE BOTTOM.

OK...

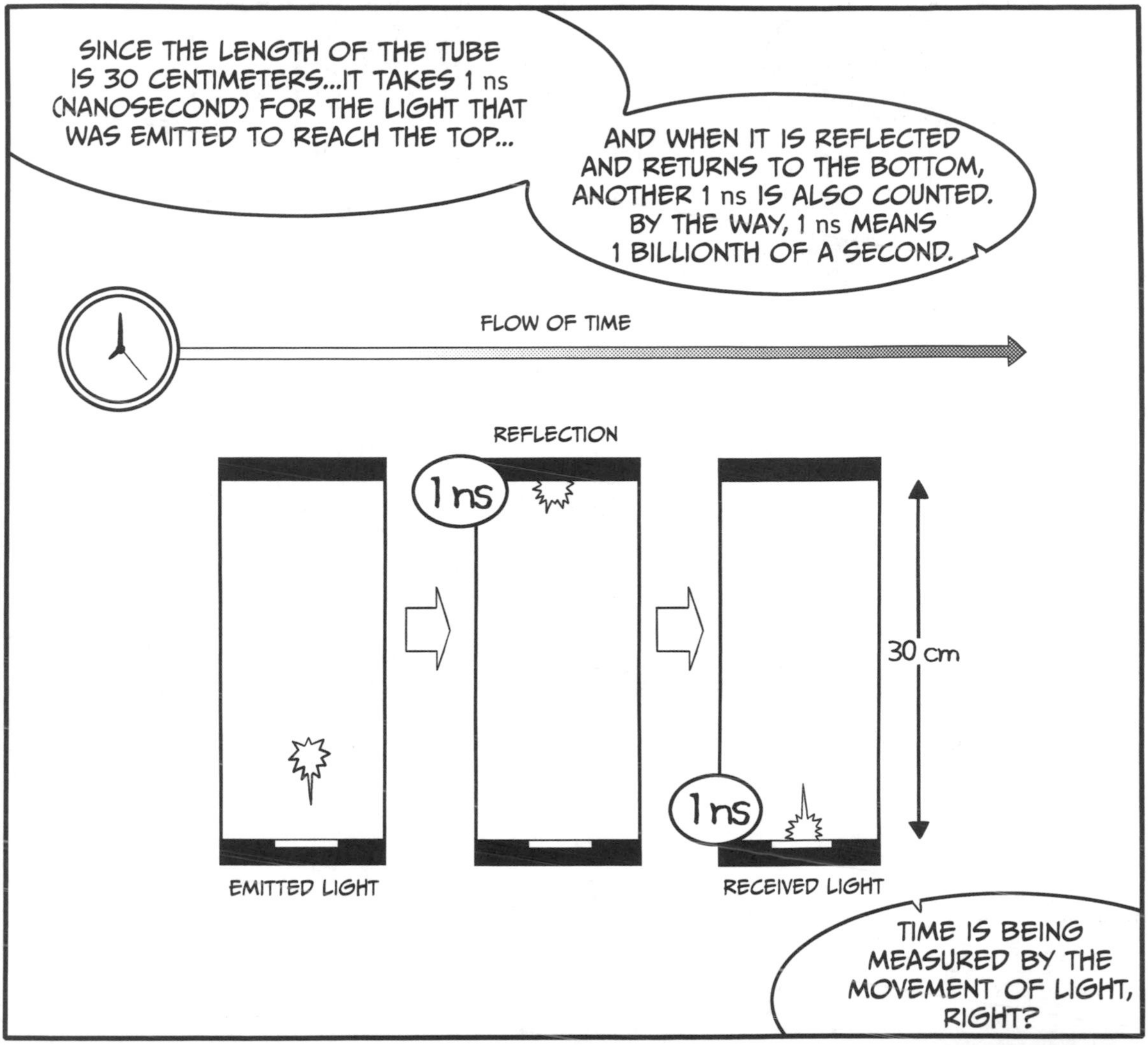
SINCE THE LENGTH OF THE TUBE IS 30 CENTIMETERS...IT TAKES 1 ns (NANOSECOND) FOR THE LIGHT THAT WAS EMITTED TO REACH THE TOP...
AND WHEN IT IS REFLECTED AND RETURNS TO THE BOTTOM, ANOTHER 1 ns IS ALSO COUNTED. BY THE WAY, 1 ns MEANS 1 BILLIONTH OF A SECOND.
FLOW OF TIME
REFLECTION
1 ns
1 ns
30 cm
EMITTED LIGHT
RECEIVED LIGHT
TIME IS BEING MEASURED BY THE MOVEMENT OF LIGHT, RIGHT?

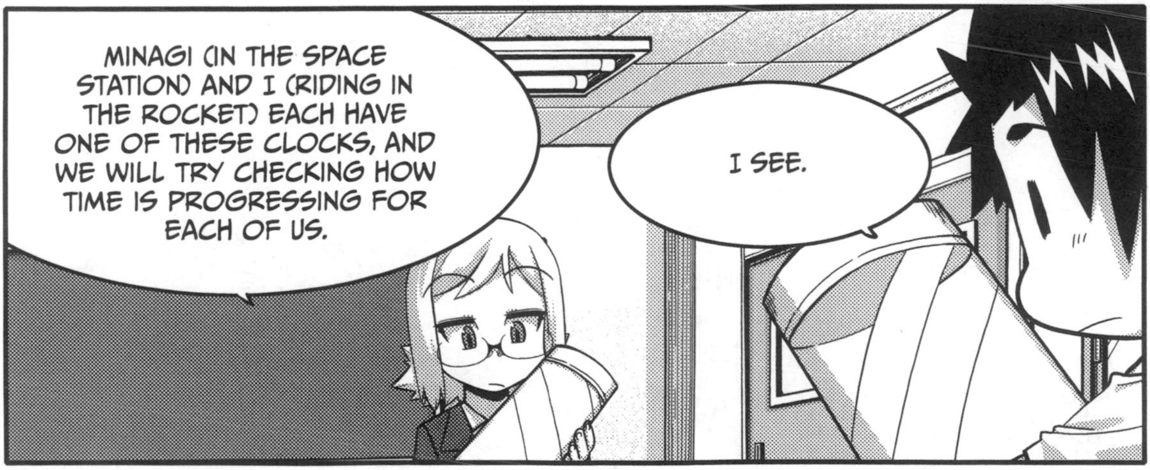
MINAGI (IN THE SPACE STATION) AND I (RIDING IN THE ROCKET) EACH HAVE ONE OF THESE CLOCKS, AND WE WILL TRY CHECKING HOW TIME IS PROGRESSING FOR EACH OF US.
I SEE.

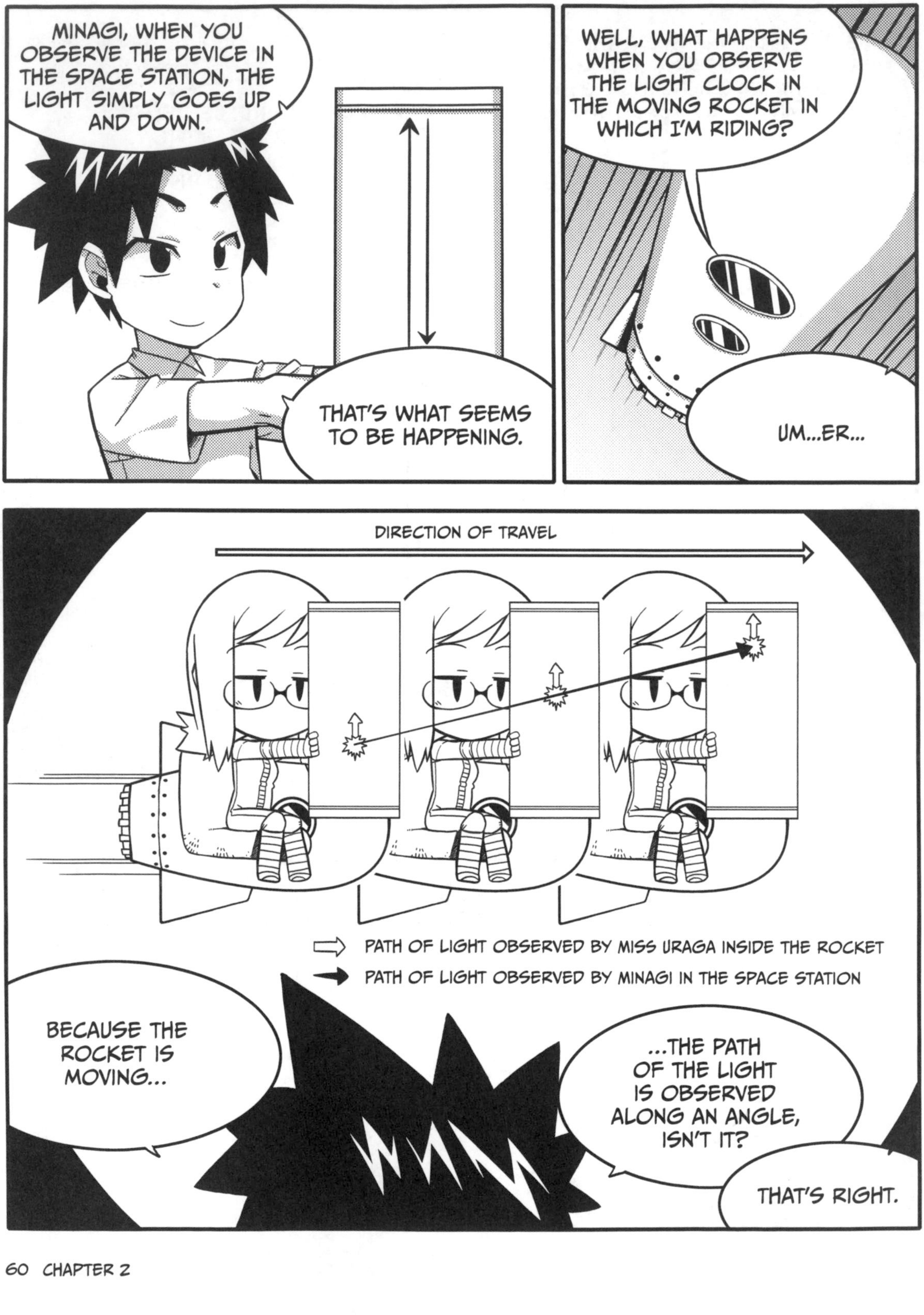
MINAGI, WHEN YOU OBSERVE THE DEVICE IN THE SPACE STATION, THE LIGHT SIMPLY GOES UP AND DOWN.
THAT'S WHAT SEEMS TO BE HAPPENING.
WELL, WHAT HAPPENS WHEN YOU OBSERVE THE LIGHT CLOCK IN THE MOVING ROCKET IN WHICH I'M RIDING?
UM...ER...
DIRECTION OF TRAVEL
PATH OF LIGHT OBSERVED BY MISS URAGA INSIDE THE ROCKET
PATH OF LIGHT OBSERVED BY MINAGI IN THE SPACE STATION
BECAUSE THE ROCKET IS MOVING...
...THE PATH OF THE LIGHT IS OBSERVED ALONG AN ANGLE, ISN'T IT?
THAT'S RIGHT.

MINAGI OBSERVES LIGHT IN HIS CLOCK MOVE DIRECTLY UPWARDS AND DOWNWARDS, BUT MINAGI OBSERVES THE LIGHT IN *MY* CLOCK PROCEED AT A LONGER PATH, AT AN ANGLE.

LET'S CONSIDER THIS USING THE *PYTHAGOREAN THEOREM.*

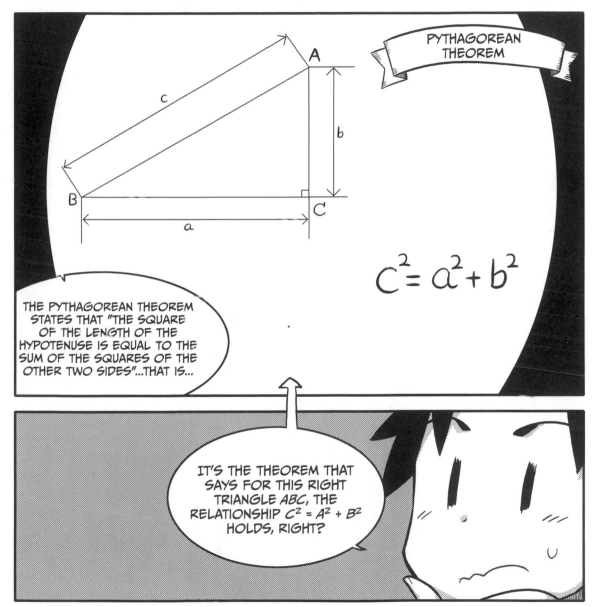

PYTHAGOREAN THEOREM

$$c^2 = a^2 + b^2$$

THE PYTHAGOREAN THEOREM STATES THAT "THE SQUARE OF THE LENGTH OF THE HYPOTENUSE IS EQUAL TO THE SUM OF THE SQUARES OF THE OTHER TWO SIDES"...THAT IS...

IT'S THE THEOREM THAT SAYS FOR THIS RIGHT TRIANGLE *ABC*, THE RELATIONSHIP $C^2 = A^2 + B^2$ HOLDS, RIGHT?

IF WE APPLY THE PYTHAGOREAN THEOREM TO THE OBSERVATIONS OF THE LIGHT CLOCKS, WE FIND THAT THE LIGHT MOVING ALONG THE HYPOTENEUSE TRAVELS A LONGER DISTANCE THAN THE HEIGHT OF THE LIGHT CLOCK, RIGHT?

YEAH, THAT'S RIGHT!

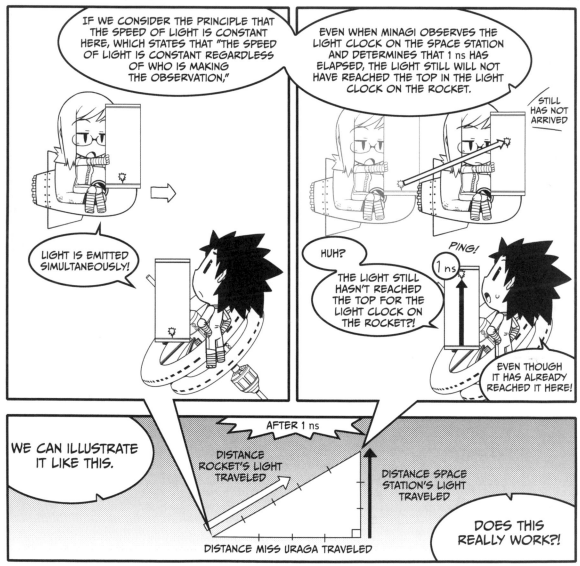

IF WE CONSIDER THE PRINCIPLE THAT THE SPEED OF LIGHT IS CONSTANT HERE, WHICH STATES THAT "THE SPEED OF LIGHT IS CONSTANT REGARDLESS OF WHO IS MAKING THE OBSERVATION,"

LIGHT IS EMITTED SIMULTANEOUSLY!

EVEN WHEN MINAGI OBSERVES THE LIGHT CLOCK ON THE SPACE STATION AND DETERMINES THAT 1 ns HAS ELAPSED, THE LIGHT STILL WILL NOT HAVE REACHED THE TOP IN THE LIGHT CLOCK ON THE ROCKET.

STILL HAS NOT ARRIVED

HUH?

THE LIGHT STILL HASN'T REACHED THE TOP FOR THE LIGHT CLOCK ON THE ROCKET?!

PING!

1 ns

EVEN THOUGH IT HAS ALREADY REACHED IT HERE!

WE CAN ILLUSTRATE IT LIKE THIS.

AFTER 1 ns

DISTANCE ROCKET'S LIGHT TRAVELED

DISTANCE SPACE STATION'S LIGHT TRAVELED

DISTANCE MISS URAGA TRAVELED

DOES THIS REALLY WORK?!

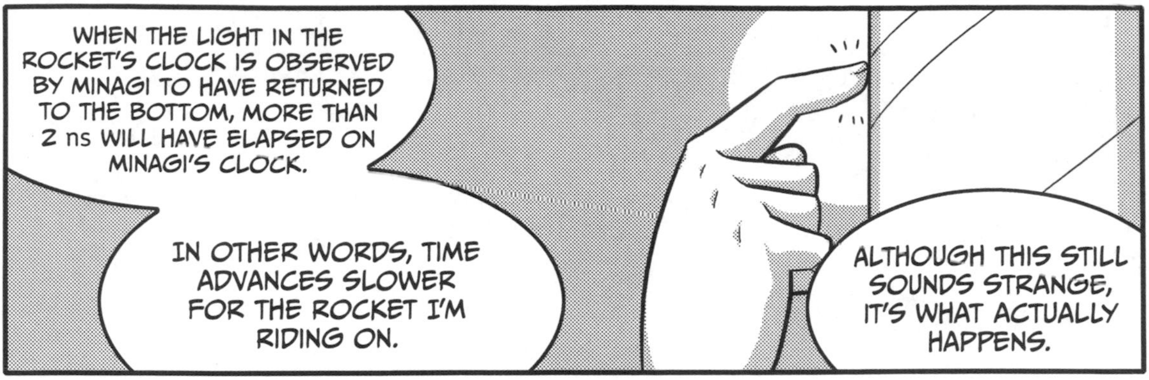
WHEN THE LIGHT IN THE ROCKET'S CLOCK IS OBSERVED BY MINAGI TO HAVE RETURNED TO THE BOTTOM, MORE THAN 2 ns WILL HAVE ELAPSED ON MINAGI'S CLOCK.
IN OTHER WORDS, TIME ADVANCES SLOWER FOR THE ROCKET I'M RIDING ON.
ALTHOUGH THIS STILL SOUNDS STRANGE, IT'S WHAT ACTUALLY HAPPENS.

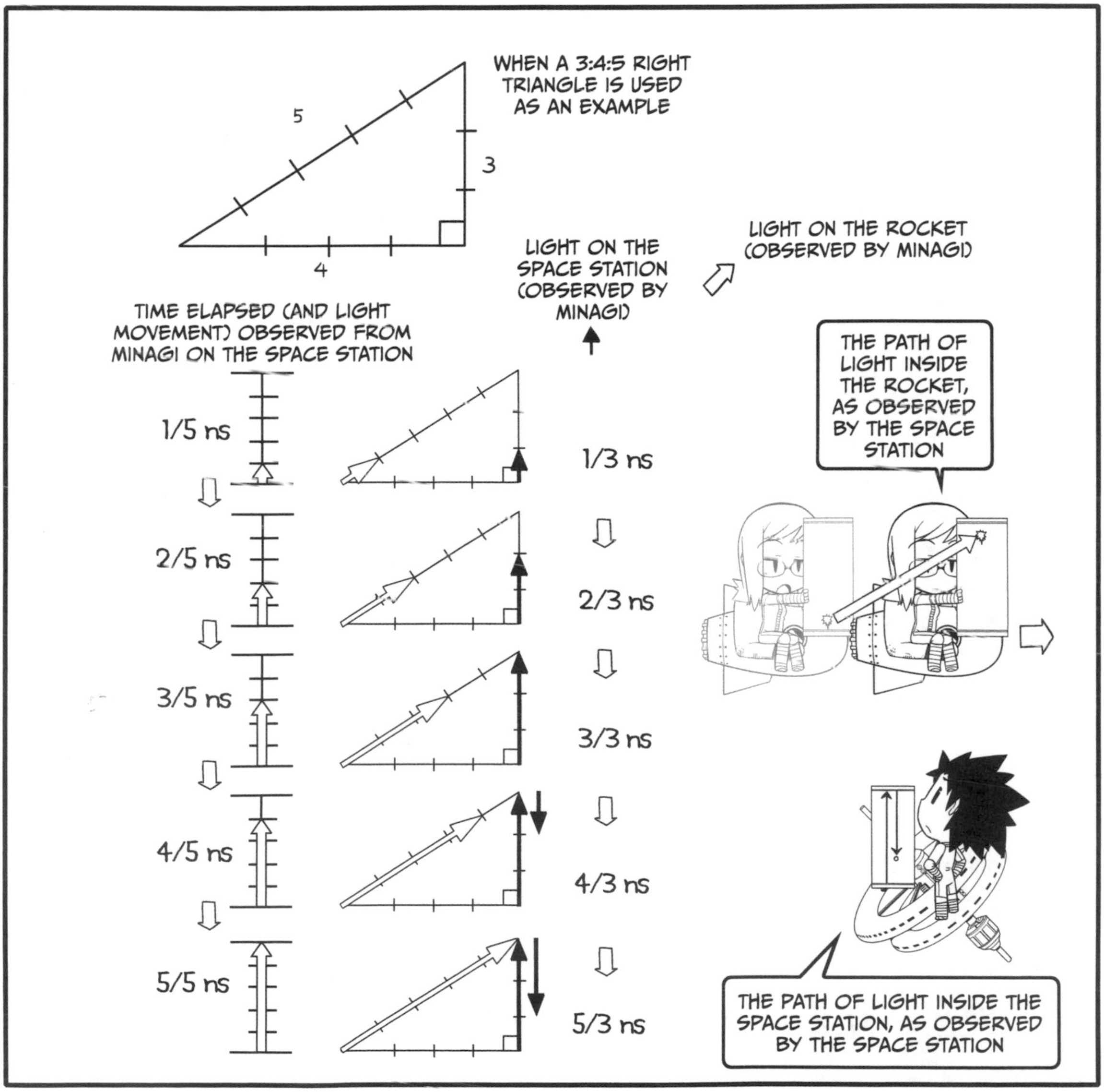
WHEN A 3:4:5 RIGHT TRIANGLE IS USED AS AN EXAMPLE
5
3
4
TIME ELAPSED (AND LIGHT MOVEMENT) OBSERVED FROM MINAGI ON THE SPACE STATION
LIGHT ON THE SPACE STATION (OBSERVED BY MINAGI)
LIGHT ON THE ROCKET (OBSERVED BY MINAGI)
1/5 ns
2/5 ns
3/5 ns
4/5 ns
5/5 ns
1/3 ns
2/3 ns
3/3 ns
4/3 ns
5/3 ns
THE PATH OF LIGHT INSIDE THE ROCKET, AS OBSERVED BY THE SPACE STATION
THE PATH OF LIGHT INSIDE THE SPACE STATION, AS OBSERVED BY THE SPACE STATION

3. THE SLOWING OF TIME MUTUALLY AFFECTS EACH PARTY EQUALLY

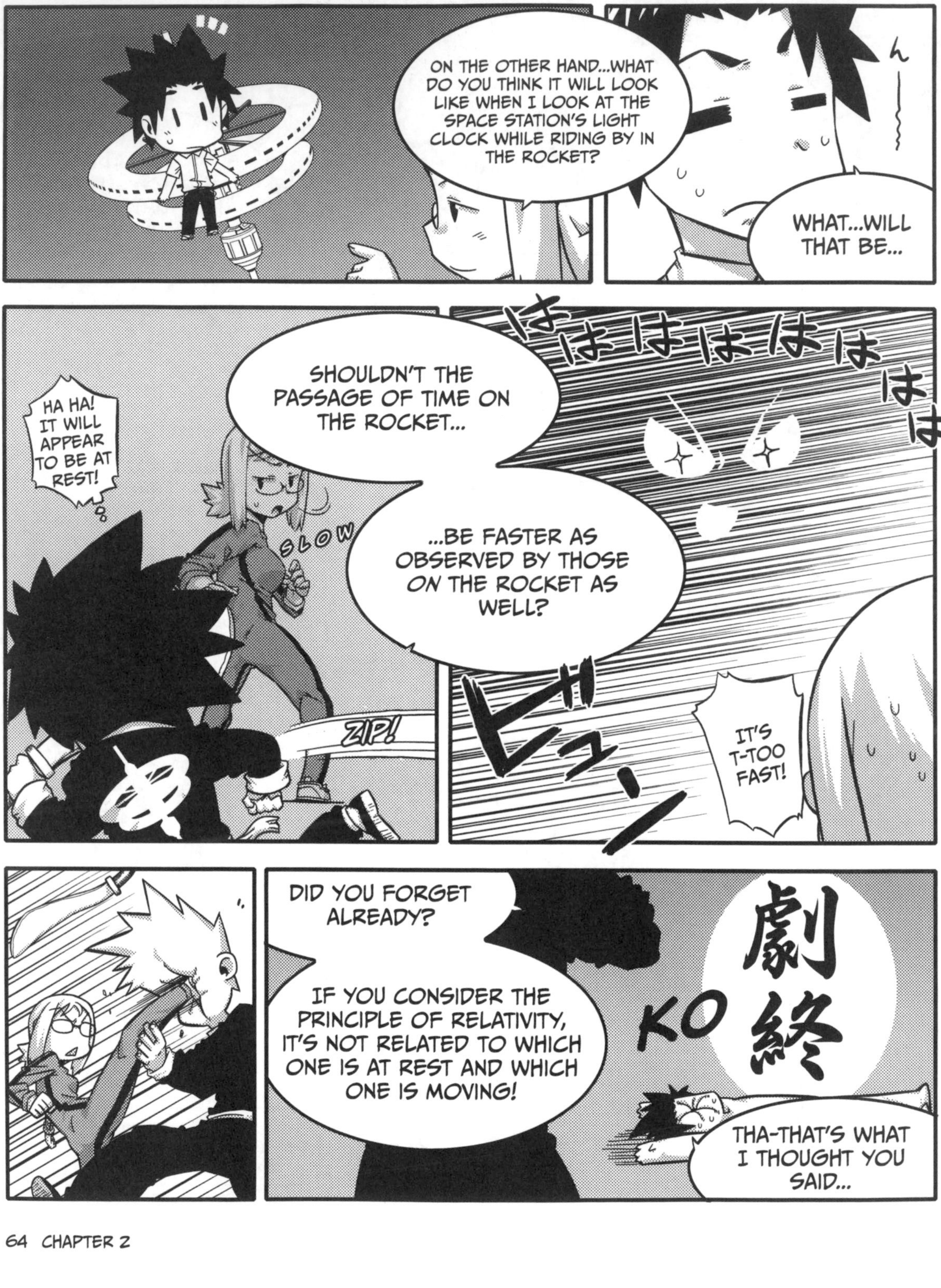
ON THE OTHER HAND...WHAT DO YOU THINK IT WILL LOOK LIKE WHEN I LOOK AT THE SPACE STATION'S LIGHT CLOCK WHILE RIDING BY IN THE ROCKET?
WHAT...WILL THAT BE...
HA HA! IT WILL APPEAR TO BE AT REST!
SHOULDN'T THE PASSAGE OF TIME ON THE ROCKET...
SLOW
...BE FASTER AS OBSERVED BY THOSE ON THE ROCKET AS WELL?
ZIP!
IT'S T-TOO FAST!
DID YOU FORGET ALREADY?
IF YOU CONSIDER THE PRINCIPLE OF RELATIVITY, IT'S NOT RELATED TO WHICH ONE IS AT REST AND WHICH ONE IS MOVING!
劇終
KO
THA-THAT'S WHAT I THOUGHT YOU SAID...

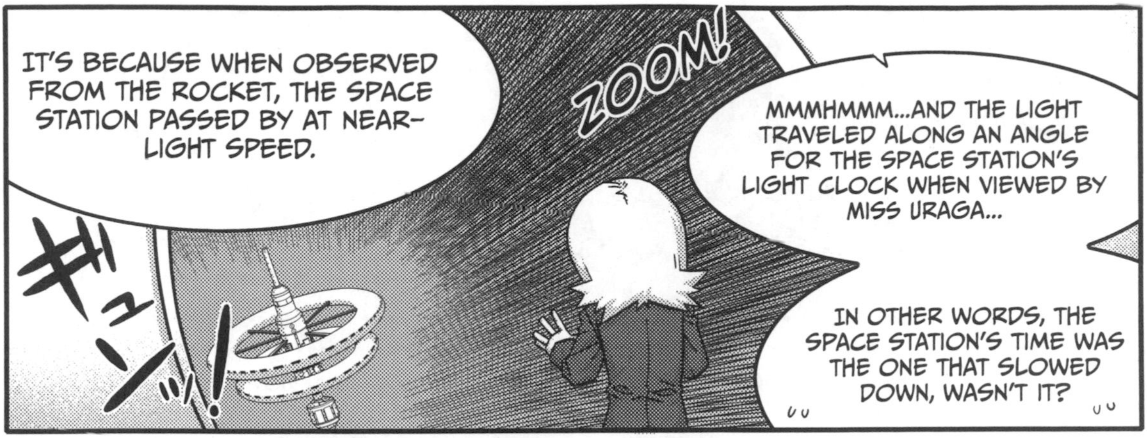
IT'S BECAUSE WHEN OBSERVED FROM THE ROCKET, THE SPACE STATION PASSED BY AT NEAR-LIGHT SPEED.
ZOOM!
ギュンッ!
MMMHMMM...AND THE LIGHT TRAVELED ALONG AN ANGLE FOR THE SPACE STATION'S LIGHT CLOCK WHEN VIEWED BY MISS URAGA...
IN OTHER WORDS, THE SPACE STATION'S TIME WAS THE ONE THAT SLOWED DOWN, WASN'T IT?

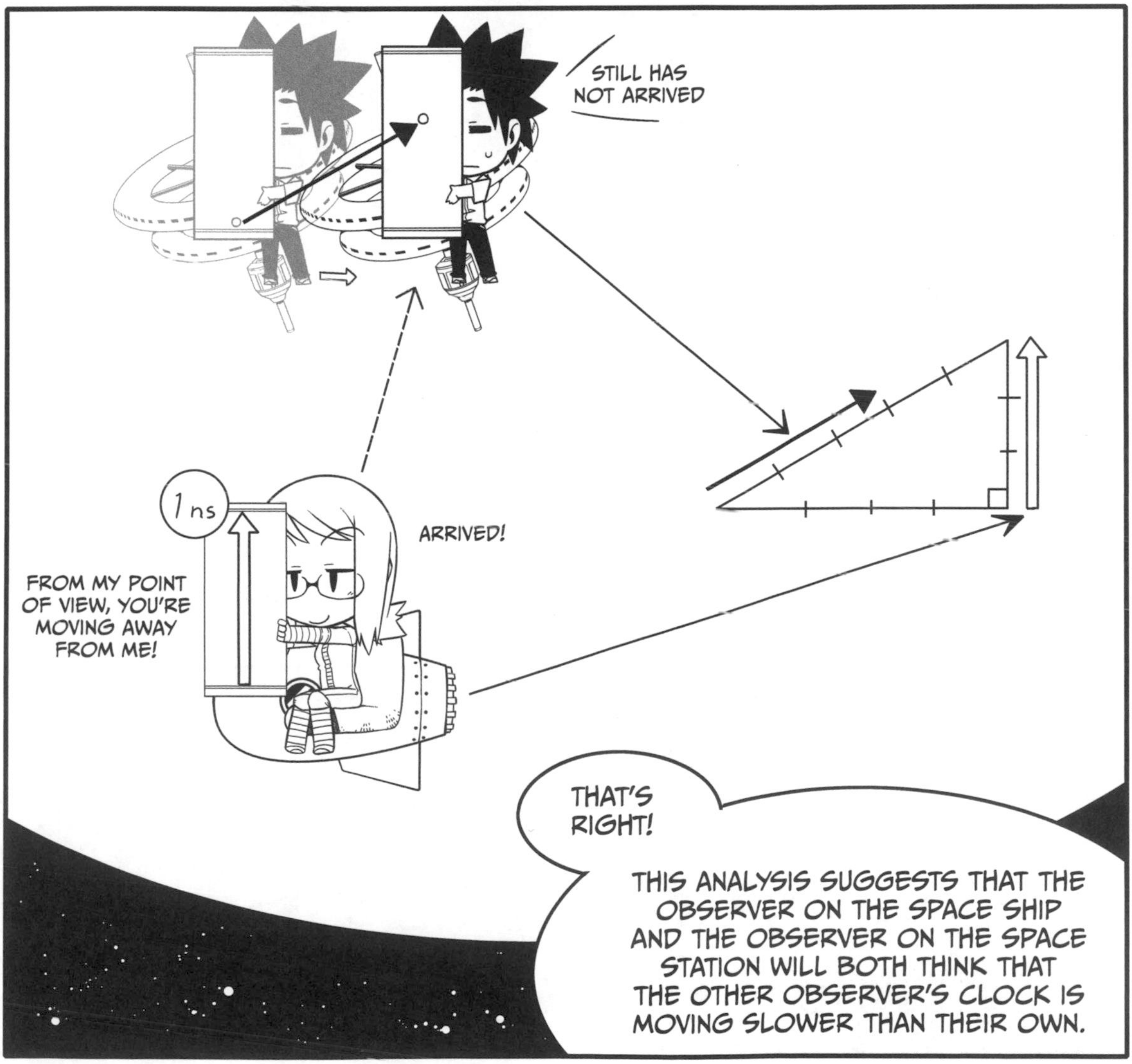
STILL HAS NOT ARRIVED
1 ns
ARRIVED!
FROM MY POINT OF VIEW, YOU'RE MOVING AWAY FROM ME!
THAT'S RIGHT!
THIS ANALYSIS SUGGESTS THAT THE OBSERVER ON THE SPACE SHIP AND THE OBSERVER ON THE SPACE STATION WILL BOTH THINK THAT THE OTHER OBSERVER'S CLOCK IS MOVING SLOWER THAN THEIR OWN.

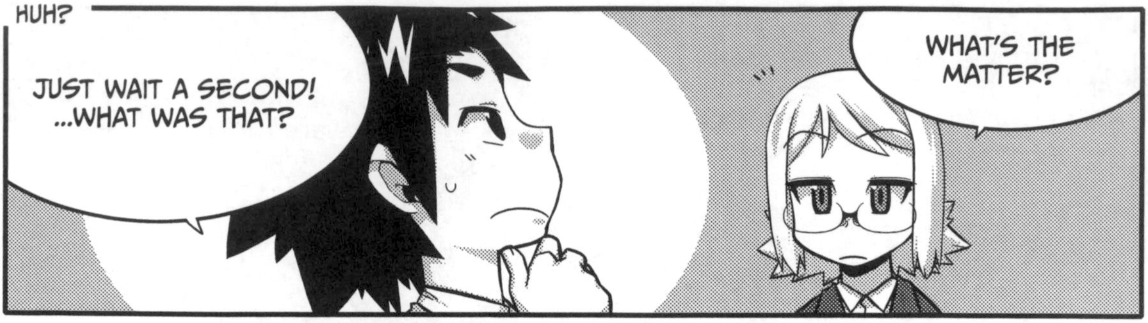
HUH?
JUST WAIT A SECOND! ...WHAT WAS THAT?
WHAT'S THE MATTER?

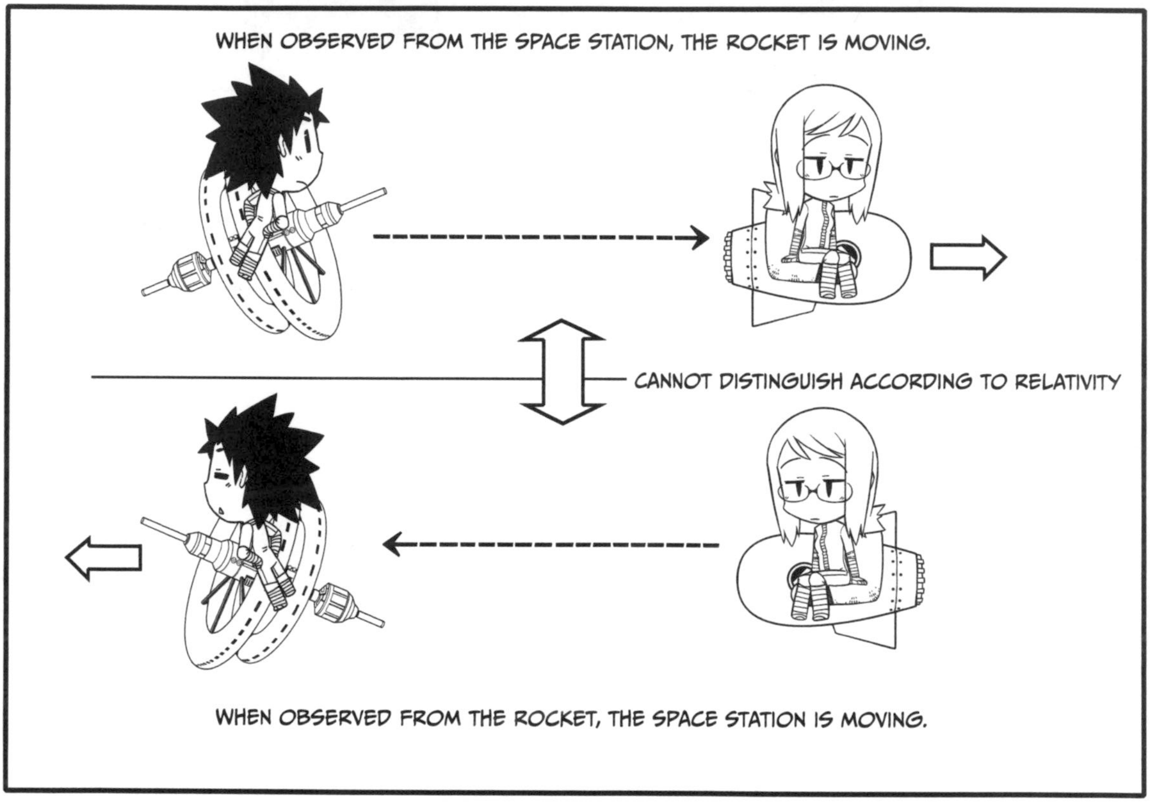
WHEN OBSERVED FROM THE SPACE STATION, THE ROCKET IS MOVING.
CANNOT DISTINGUISH ACCORDING TO RELATIVITY
WHEN OBSERVED FROM THE ROCKET, THE SPACE STATION IS MOVING.

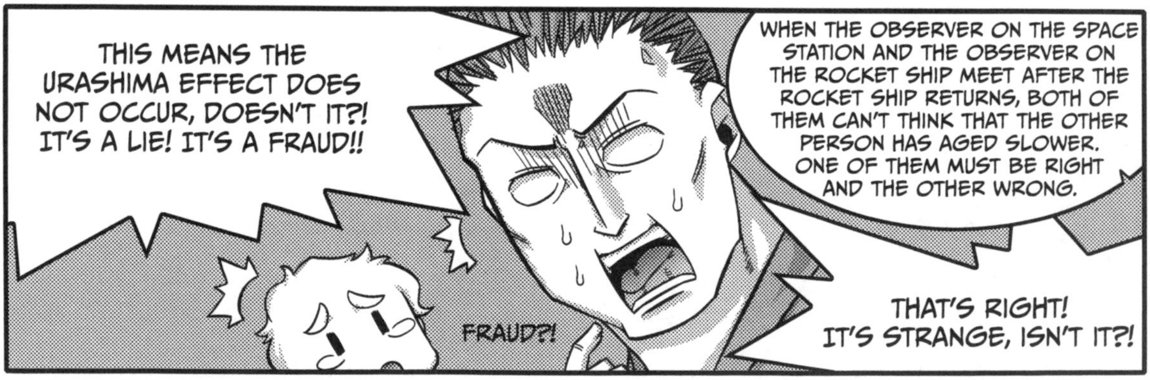
THIS MEANS THE URASHIMA EFFECT DOES NOT OCCUR, DOESN'T IT?! IT'S A LIE! IT'S A FRAUD!!
WHEN THE OBSERVER ON THE SPACE STATION AND THE OBSERVER ON THE ROCKET SHIP MEET AFTER THE ROCKET SHIP RETURNS, BOTH OF THEM CAN'T THINK THAT THE OTHER PERSON HAS AGED SLOWER. ONE OF THEM MUST BE RIGHT AND THE OTHER WRONG.
FRAUD?!
THAT'S RIGHT! IT'S STRANGE, ISN'T IT?!

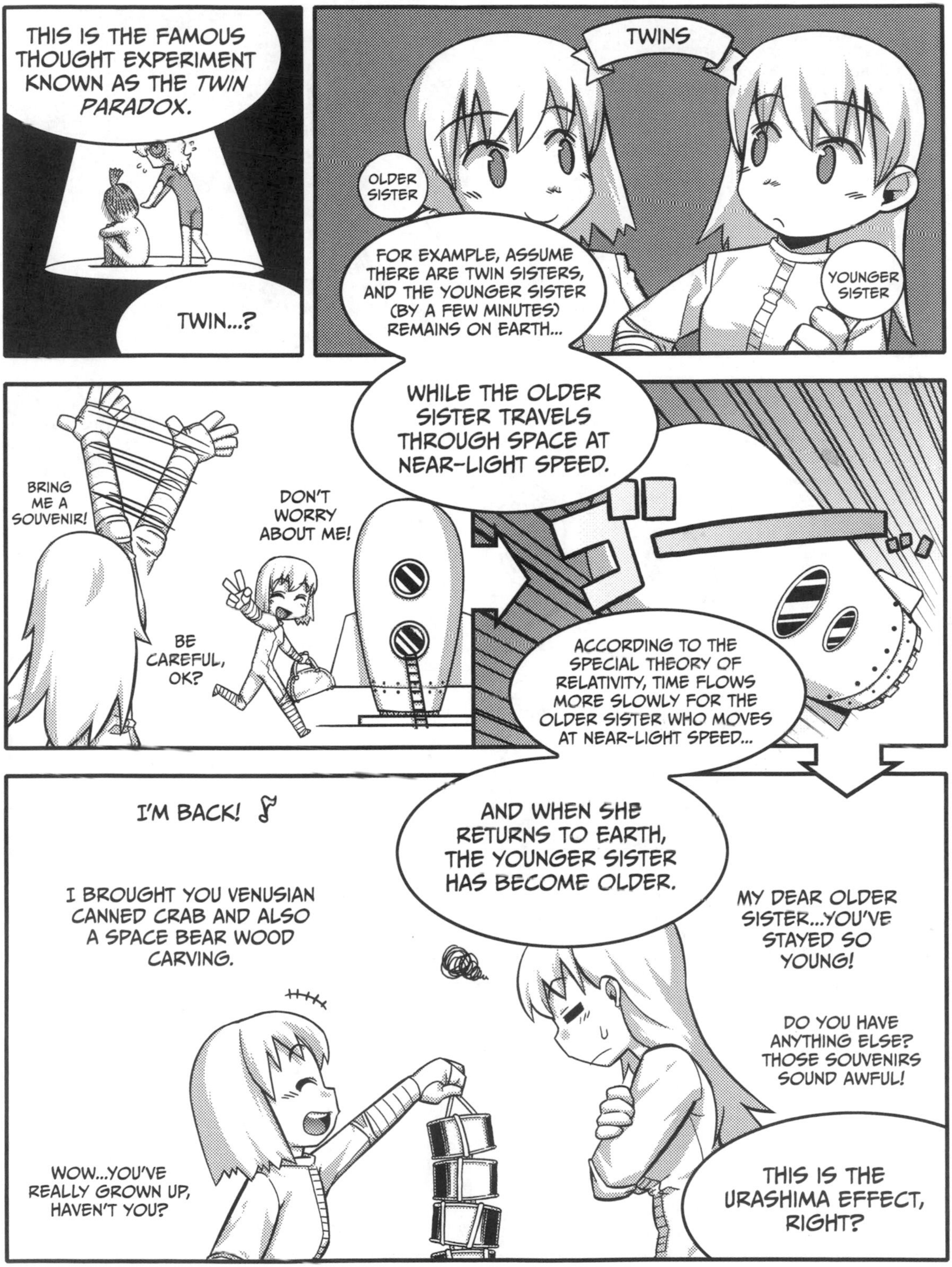
THIS IS THE FAMOUS THOUGHT EXPERIMENT KNOWN AS THE *TWIN PARADOX*.
TWIN...?
TWINS
OLDER SISTER
YOUNGER SISTER
FOR EXAMPLE, ASSUME THERE ARE TWIN SISTERS, AND THE YOUNGER SISTER (BY A FEW MINUTES) REMAINS ON EARTH...
WHILE THE OLDER SISTER TRAVELS THROUGH SPACE AT NEAR-LIGHT SPEED.
BRING ME A SOUVENIR!
DON'T WORRY ABOUT ME!
BE CAREFUL, OK?
ACCORDING TO THE SPECIAL THEORY OF RELATIVITY, TIME FLOWS MORE SLOWLY FOR THE OLDER SISTER WHO MOVES AT NEAR-LIGHT SPEED...
I'M BACK!
AND WHEN SHE RETURNS TO EARTH, THE YOUNGER SISTER HAS BECOME OLDER.
I BROUGHT YOU VENUSIAN CANNED CRAB AND ALSO A SPACE BEAR WOOD CARVING.
MY DEAR OLDER SISTER...YOU'VE STAYED SO YOUNG!
DO YOU HAVE ANYTHING ELSE? THOSE SOUVENIRS SOUND AWFUL!
WOW...YOU'VE REALLY GROWN UP, HAVEN'T YOU?
THIS IS THE URASHIMA EFFECT, RIGHT?

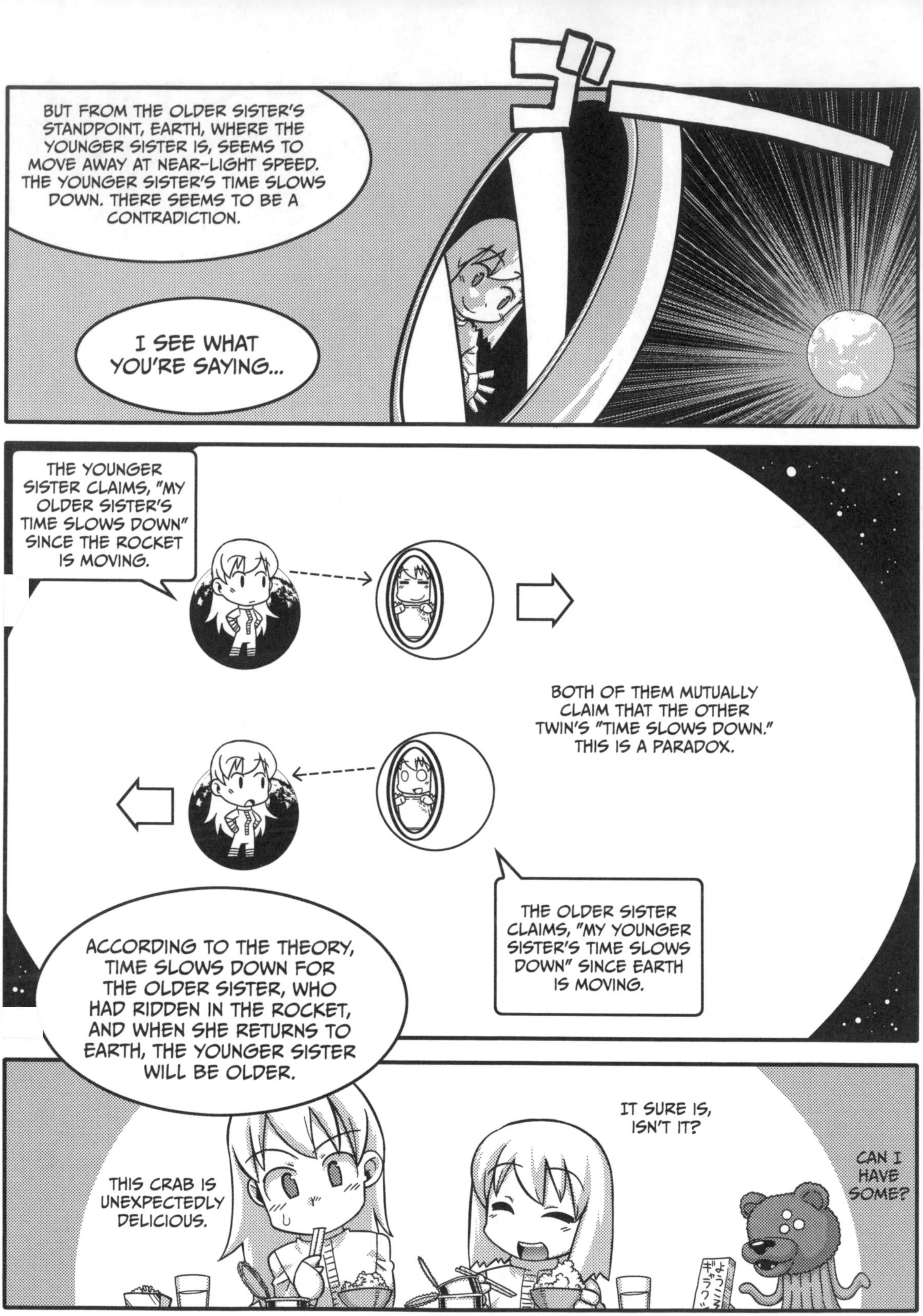
BUT FROM THE OLDER SISTER'S STANDPOINT, EARTH, WHERE THE YOUNGER SISTER IS, SEEMS TO MOVE AWAY AT NEAR-LIGHT SPEED. THE YOUNGER SISTER'S TIME SLOWS DOWN. THERE SEEMS TO BE A CONTRADICTION.
I SEE WHAT YOU'RE SAYING...
THE YOUNGER SISTER CLAIMS, "MY OLDER SISTER'S TIME SLOWS DOWN" SINCE THE ROCKET IS MOVING.
BOTH OF THEM MUTUALLY CLAIM THAT THE OTHER TWIN'S "TIME SLOWS DOWN." THIS IS A PARADOX.
THE OLDER SISTER CLAIMS, "MY YOUNGER SISTER'S TIME SLOWS DOWN" SINCE EARTH IS MOVING.
ACCORDING TO THE THEORY, TIME SLOWS DOWN FOR THE OLDER SISTER, WHO HAD RIDDEN IN THE ROCKET, AND WHEN SHE RETURNS TO EARTH, THE YOUNGER SISTER WILL BE OLDER.
THIS CRAB IS UNEXPECTEDLY DELICIOUS.
IT SURE IS, ISN'T IT?
CAN I HAVE SOME?

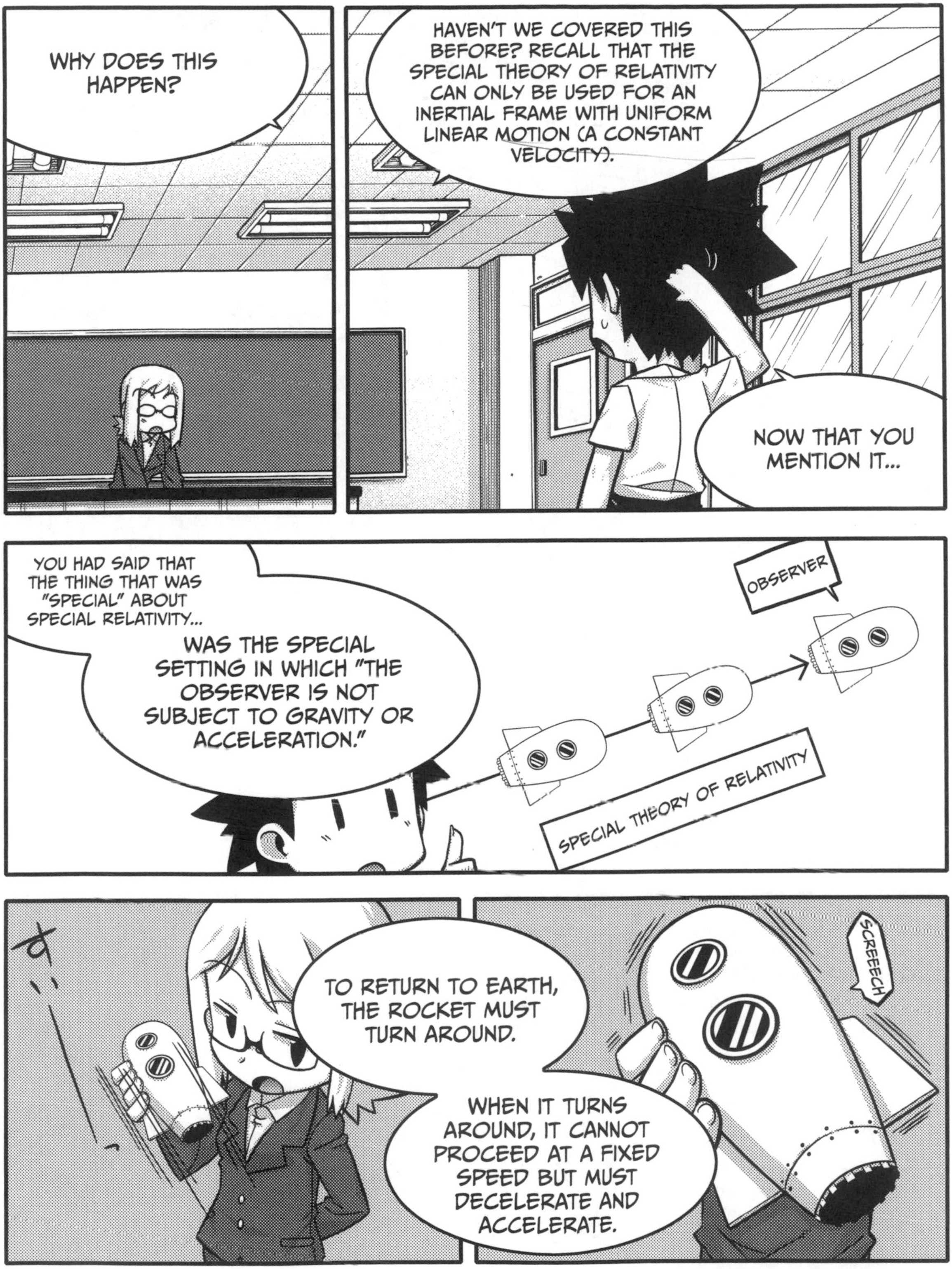
WHY DOES THIS HAPPEN?
HAVEN'T WE COVERED THIS BEFORE? RECALL THAT THE SPECIAL THEORY OF RELATIVITY CAN ONLY BE USED FOR AN INERTIAL FRAME WITH UNIFORM LINEAR MOTION (A CONSTANT VELOCITY).
NOW THAT YOU MENTION IT...
YOU HAD SAID THAT THE THING THAT WAS "SPECIAL" ABOUT SPECIAL RELATIVITY...
WAS THE SPECIAL SETTING IN WHICH "THE OBSERVER IS NOT SUBJECT TO GRAVITY OR ACCELERATION."
OBSERVER
SPECIAL THEORY OF RELATIVITY
TO RETURN TO EARTH, THE ROCKET MUST TURN AROUND.
WHEN IT TURNS AROUND, IT CANNOT PROCEED AT A FIXED SPEED BUT MUST DECELERATE AND ACCELERATE.
SCREEECH

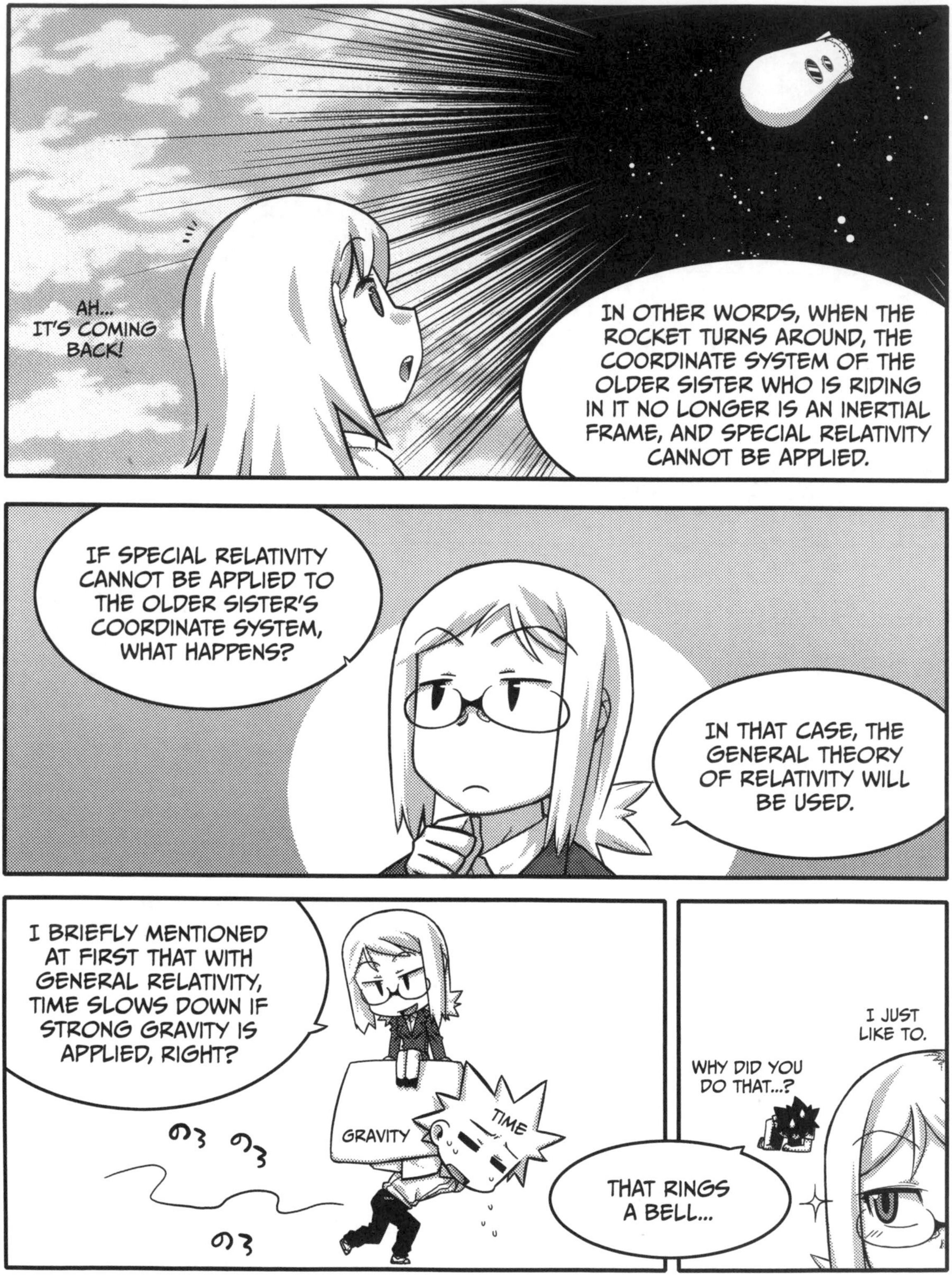
AH... IT'S COMING BACK!
IN OTHER WORDS, WHEN THE ROCKET TURNS AROUND, THE COORDINATE SYSTEM OF THE OLDER SISTER WHO IS RIDING IN IT NO LONGER IS AN INERTIAL FRAME, AND SPECIAL RELATIVITY CANNOT BE APPLIED.
IF SPECIAL RELATIVITY CANNOT BE APPLIED TO THE OLDER SISTER'S COORDINATE SYSTEM, WHAT HAPPENS?
IN THAT CASE, THE GENERAL THEORY OF RELATIVITY WILL BE USED.
I BRIEFLY MENTIONED AT FIRST THAT WITH GENERAL RELATIVITY, TIME SLOWS DOWN IF STRONG GRAVITY IS APPLIED, RIGHT?
TIME
GRAVITY
のろ のろ
のろ
THAT RINGS A BELL...
WHY DID YOU DO THAT...?
I JUST LIKE TO.

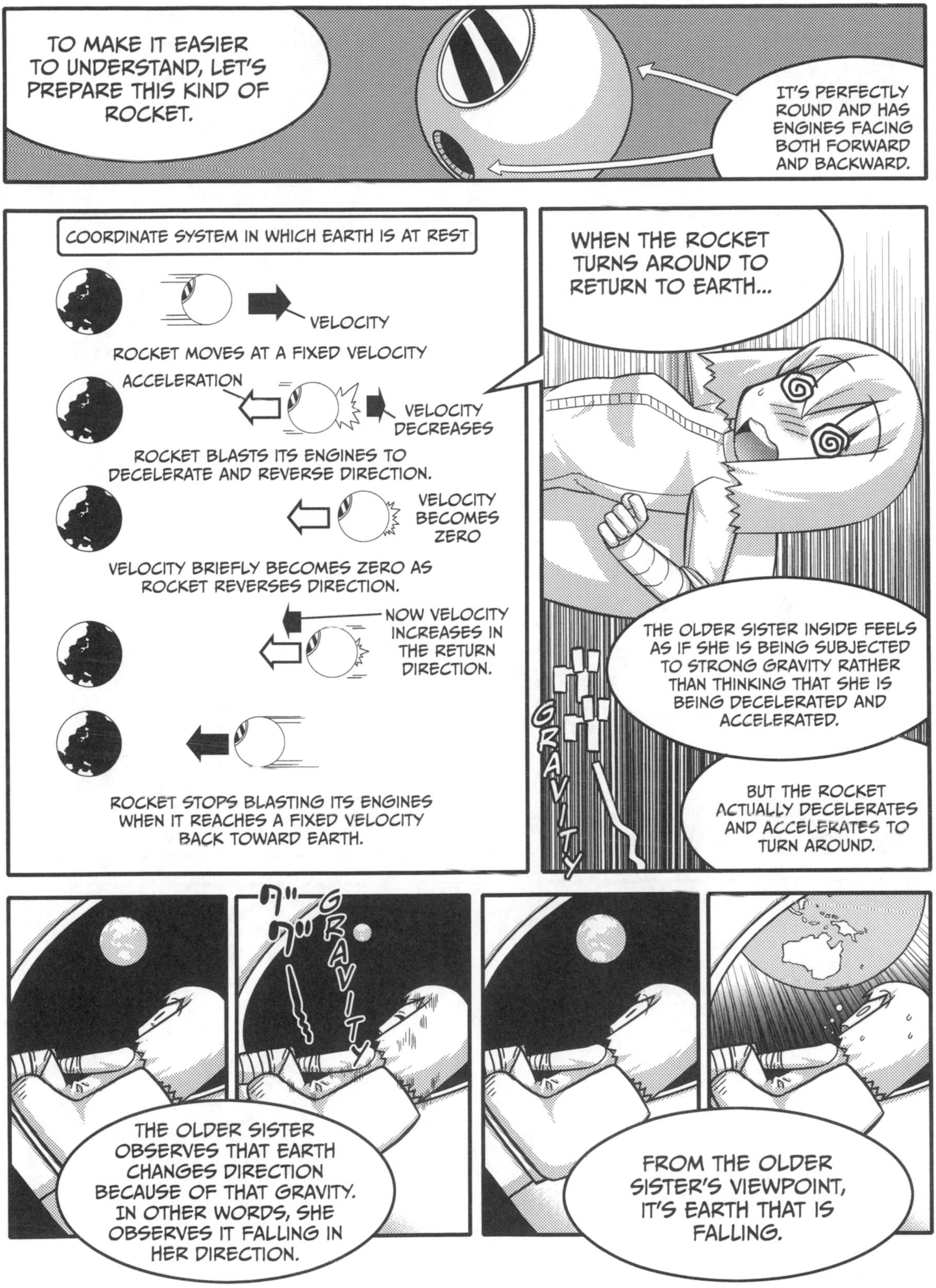

TO MAKE IT EASIER TO UNDERSTAND, LET'S PREPARE THIS KIND OF ROCKET.

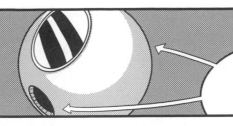

IT'S PERFECTLY ROUND AND HAS ENGINES FACING BOTH FORWARD AND BACKWARD.

COORDINATE SYSTEM IN WHICH EARTH IS AT REST

   VELOCITY

ROCKET MOVES AT A FIXED VELOCITY

ACCELERATION

   VELOCITY DECREASES

ROCKET BLASTS ITS ENGINES TO DECELERATE AND REVERSE DIRECTION.

  VELOCITY BECOMES ZERO

VELOCITY BRIEFLY BECOMES ZERO AS ROCKET REVERSES DIRECTION.

  NOW VELOCITY INCREASES IN THE RETURN DIRECTION.

ROCKET STOPS BLASTING ITS ENGINES WHEN IT REACHES A FIXED VELOCITY BACK TOWARD EARTH.

WHEN THE ROCKET TURNS AROUND TO RETURN TO EARTH...

GRAVITY

THE OLDER SISTER INSIDE FEELS AS IF SHE IS BEING SUBJECTED TO STRONG GRAVITY RATHER THAN THINKING THAT SHE IS BEING DECELERATED AND ACCELERATED.

BUT THE ROCKET ACTUALLY DECELERATES AND ACCELERATES TO TURN AROUND.

GRAVITY

THE OLDER SISTER OBSERVES THAT EARTH CHANGES DIRECTION BECAUSE OF THAT GRAVITY. IN OTHER WORDS, SHE OBSERVES IT FALLING IN HER DIRECTION.

FROM THE OLDER SISTER'S VIEWPOINT, IT'S EARTH THAT IS FALLING.

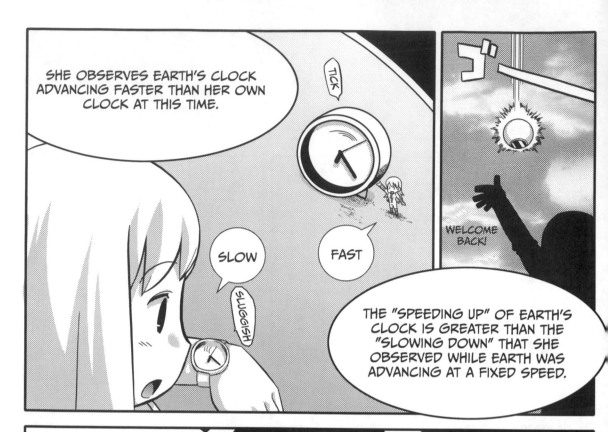

SHE OBSERVES EARTH'S CLOCK ADVANCING FASTER THAN HER OWN CLOCK AT THIS TIME.

TICK

SLOW

SLUGGISH

FAST

WELCOME BACK!

THE "SPEEDING UP" OF EARTH'S CLOCK IS GREATER THAN THE "SLOWING DOWN" THAT SHE OBSERVED WHILE EARTH WAS ADVANCING AT A FIXED SPEED.

THE ROCKET'S ACCELERATION IS ALSO *GRAVITY*?

ALTHOUGH I'M GOING TO OMIT A DIFFICULT EXPLANATION ABOUT THAT NOW...

...THE *EQUIVALENCE PRINCIPLE* STATES THAT THE APPARENT FORCE THAT A PERSON IN THE ROCKET IS SUBJECTED TO DUE TO THE ROCKET'S ACCELERATION IS CONSIDERED TO BE THE SAME AS GRAVITY.

THAT WAS A FOUNDATIONAL PRINCIPLE EINSTEIN USED WHEN HE CREATED THE GENERAL THEORY OF RELATIVITY.

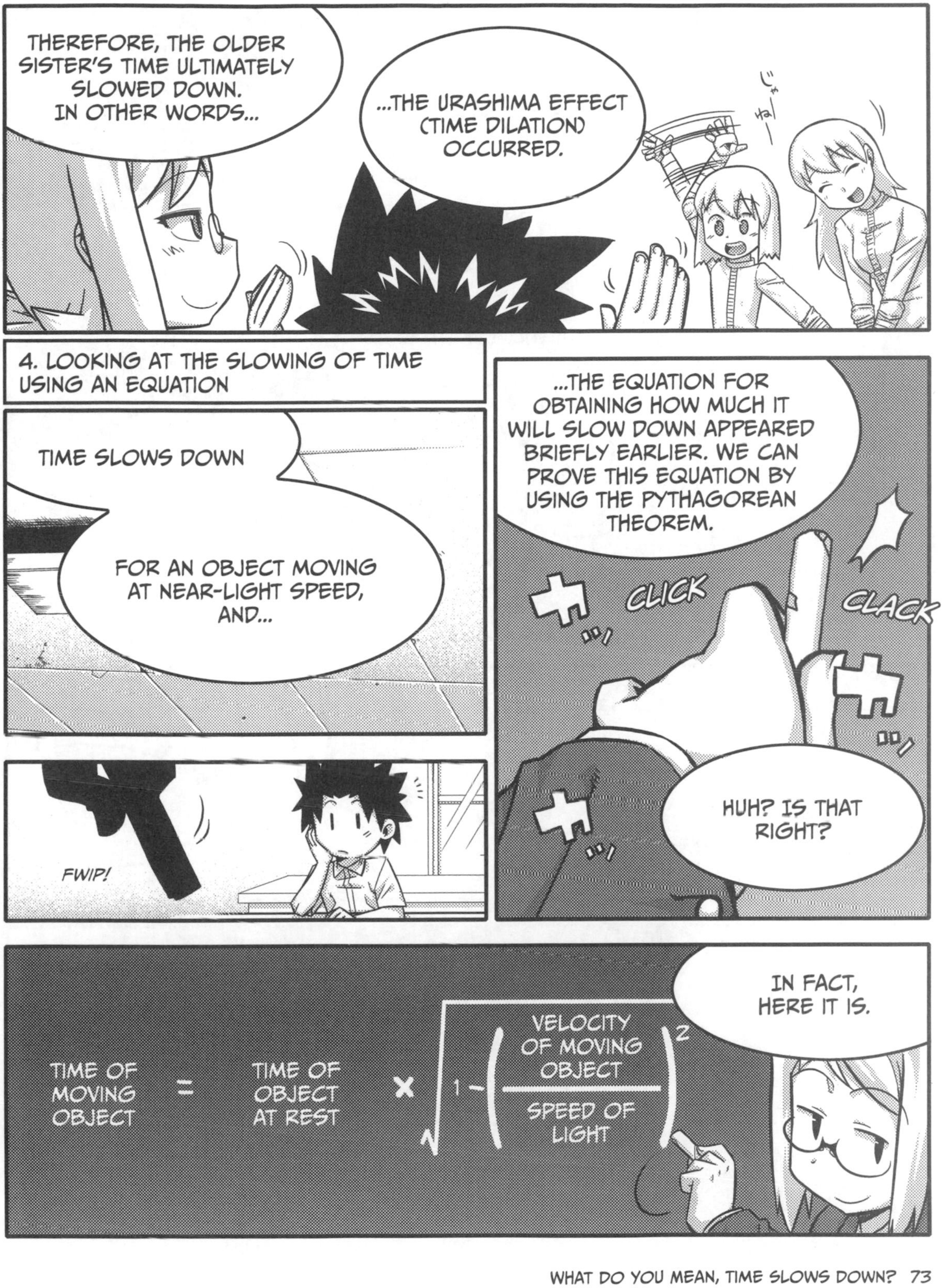

THEREFORE, THE OLDER SISTER'S TIME ULTIMATELY SLOWED DOWN. IN OTHER WORDS...
...THE URASHIMA EFFECT (TIME DILATION) OCCURRED.
4. LOOKING AT THE SLOWING OF TIME USING AN EQUATION
TIME SLOWS DOWN
FOR AN OBJECT MOVING AT NEAR-LIGHT SPEED, AND...
FWIP!
...THE EQUATION FOR OBTAINING HOW MUCH IT WILL SLOW DOWN APPEARED BRIEFLY EARLIER. WE CAN PROVE THIS EQUATION BY USING THE PYTHAGOREAN THEOREM.
CLICK
CLACK
HUH? IS THAT RIGHT?
IN FACT, HERE IT IS.
TIME OF MOVING OBJECT = TIME OF OBJECT AT REST × √(1 − (VELOCITY OF MOVING OBJECT / SPEED OF LIGHT)²)

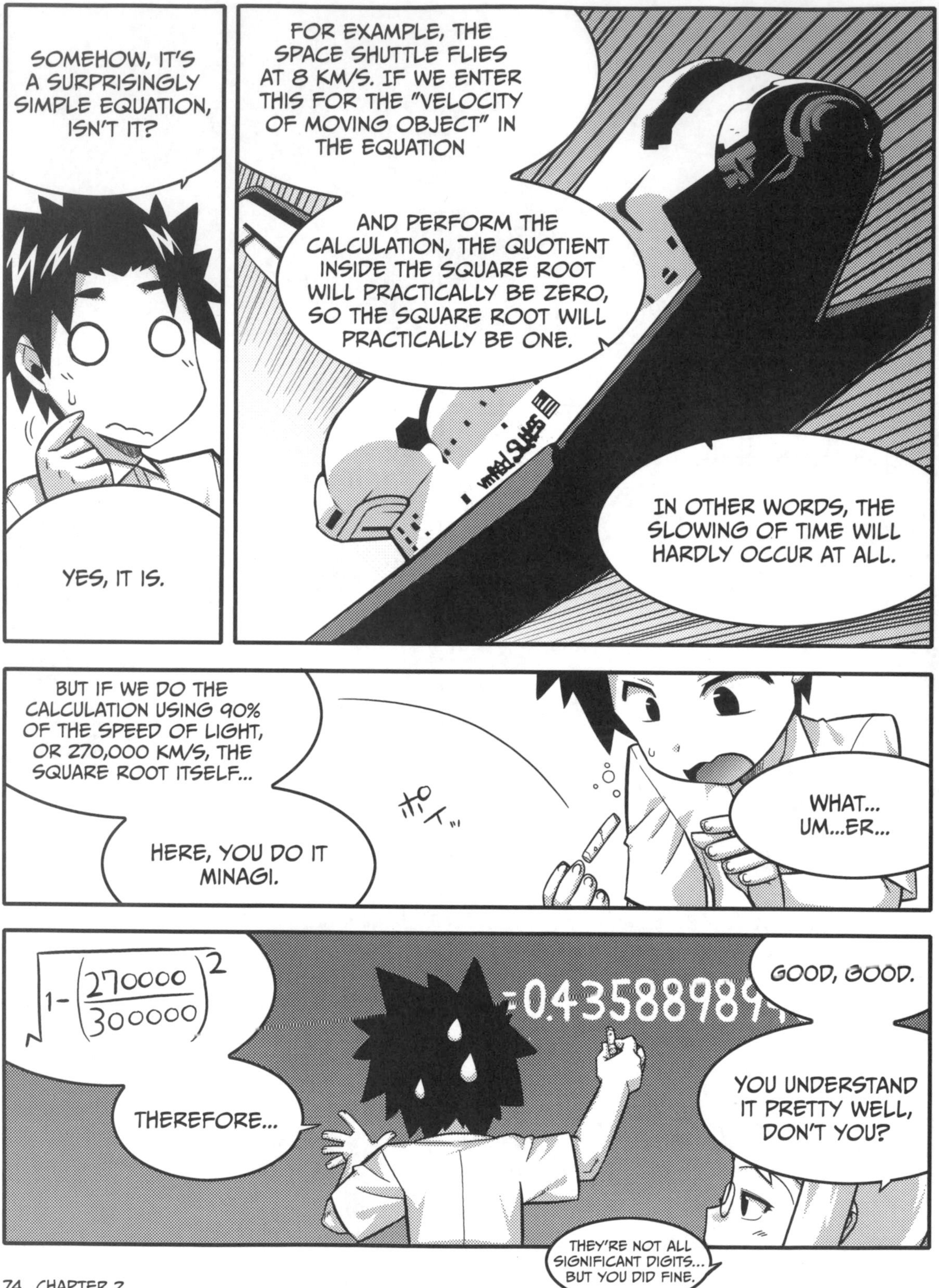
SOMEHOW, IT'S A SURPRISINGLY SIMPLE EQUATION, ISN'T IT?
YES, IT IS.
FOR EXAMPLE, THE SPACE SHUTTLE FLIES AT 8 KM/S. IF WE ENTER THIS FOR THE "VELOCITY OF MOVING OBJECT" IN THE EQUATION
AND PERFORM THE CALCULATION, THE QUOTIENT INSIDE THE SQUARE ROOT WILL PRACTICALLY BE ZERO, SO THE SQUARE ROOT WILL PRACTICALLY BE ONE.
IN OTHER WORDS, THE SLOWING OF TIME WILL HARDLY OCCUR AT ALL.
BUT IF WE DO THE CALCULATION USING 90% OF THE SPEED OF LIGHT, OR 270,000 KM/S, THE SQUARE ROOT ITSELF...
HERE, YOU DO IT MINAGI.
WHAT... UM...ER...
$\sqrt{1-\left(\frac{270000}{300000}\right)^2}$
=0.43588989
THEREFORE...
GOOD, GOOD.
YOU UNDERSTAND IT PRETTY WELL, DON'T YOU?
THEY'RE NOT ALL SIGNIFICANT DIGITS... BUT YOU DID FINE.

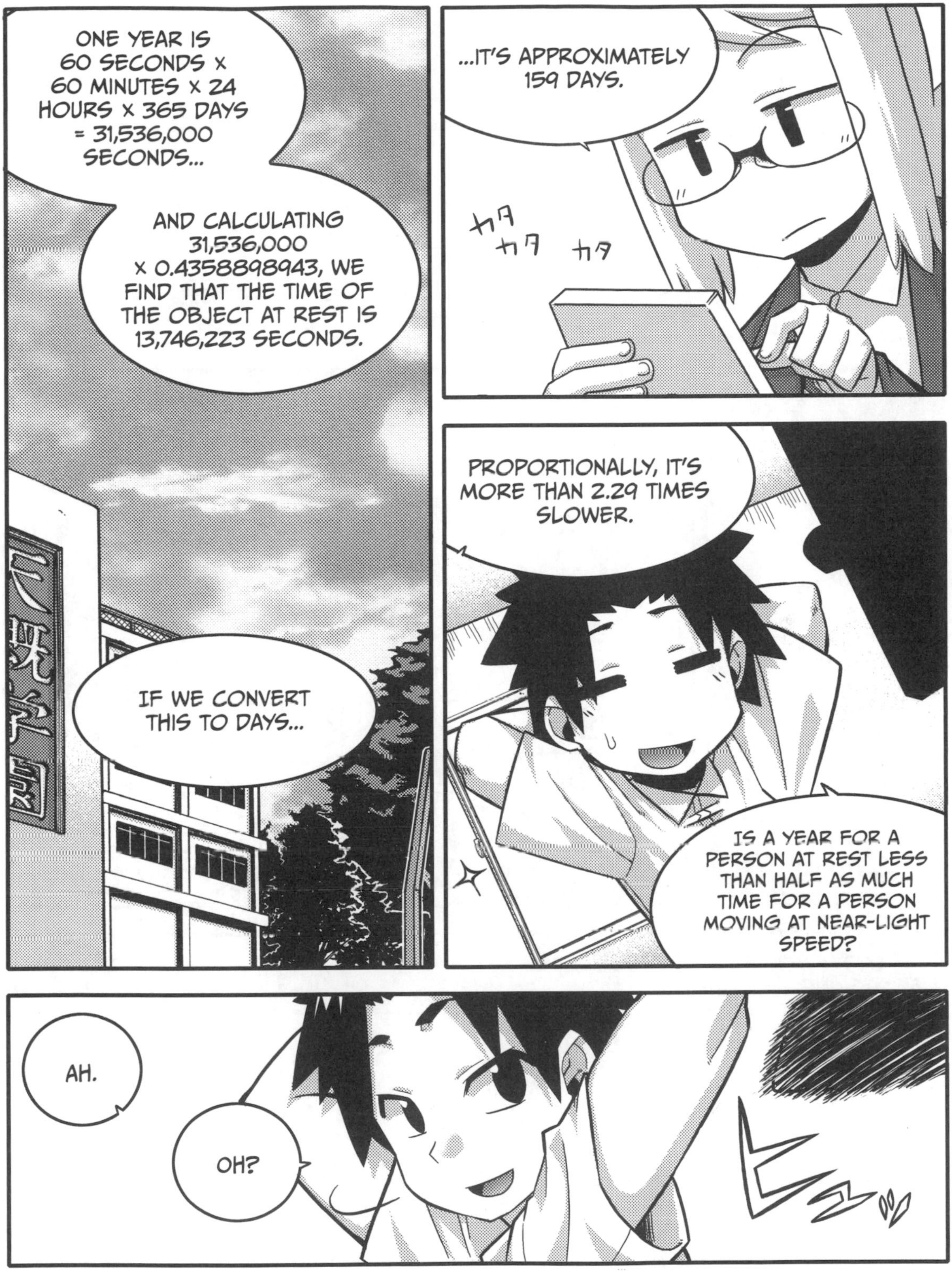
ONE YEAR IS 60 SECONDS x 60 MINUTES x 24 HOURS x 365 DAYS = 31,536,000 SECONDS...
AND CALCULATING 31,536,000 x 0.4358898943, WE FIND THAT THE TIME OF THE OBJECT AT REST IS 13,746,223 SECONDS.
IF WE CONVERT THIS TO DAYS...
...IT'S APPROXIMATELY 159 DAYS.
カタ
カタ
カタ
PROPORTIONALLY, IT'S MORE THAN 2.29 TIMES SLOWER.
IS A YEAR FOR A PERSON AT REST LESS THAN HALF AS MUCH TIME FOR A PERSON MOVING AT NEAR-LIGHT SPEED?
AH.
OH?

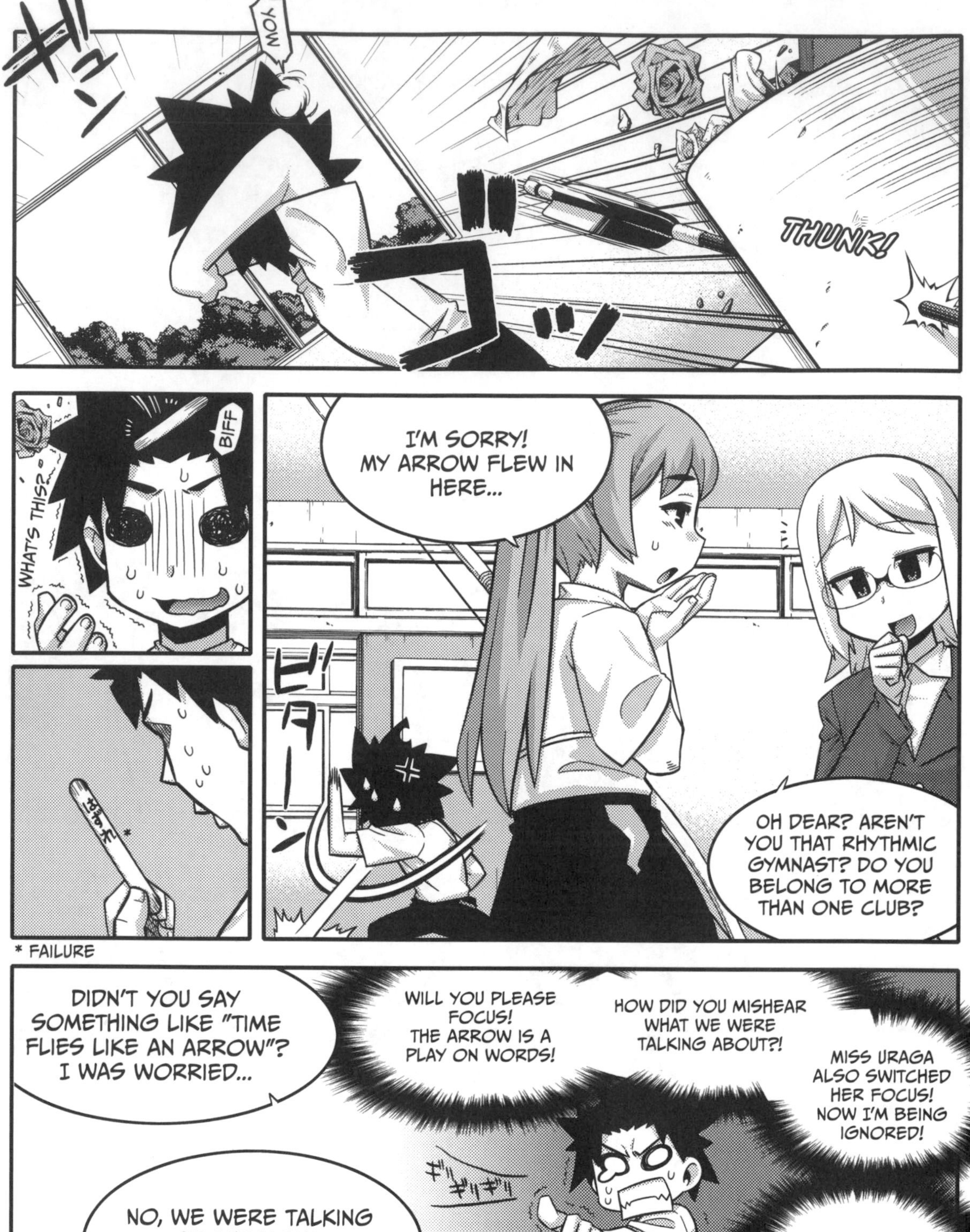
YOW
THUNK!
BIFF
WHAT'S THIS?
I'M SORRY! MY ARROW FLEW IN HERE...
OH DEAR? AREN'T YOU THAT RHYTHMIC GYMNAST? DO YOU BELONG TO MORE THAN ONE CLUB?
* FAILURE
DIDN'T YOU SAY SOMETHING LIKE "TIME FLIES LIKE AN ARROW"? I WAS WORRIED...
WILL YOU PLEASE FOCUS! THE ARROW IS A PLAY ON WORDS!
HOW DID YOU MISHEAR WHAT WE WERE TALKING ABOUT?!
MISS URAGA ALSO SWITCHED HER FOCUS! NOW I'M BEING IGNORED!
NO, WE WERE TALKING ABOUT LIGHT. HA HA HA!
DAMN, I CAN'T EVEN FOLLOW WHAT'S GOING ON!

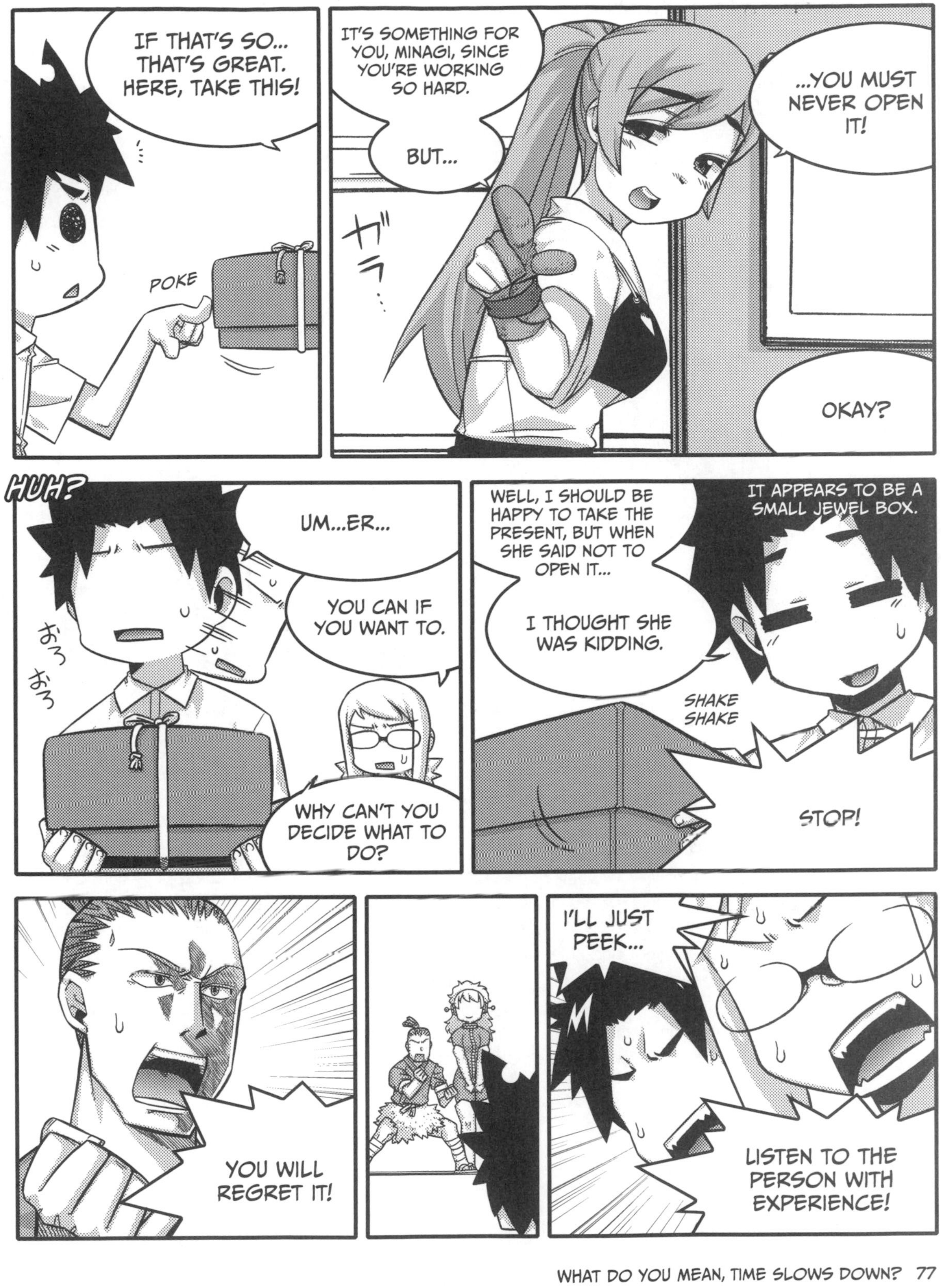
IF THAT'S SO... THAT'S GREAT. HERE, TAKE THIS!
POKE
IT'S SOMETHING FOR YOU, MINAGI, SINCE YOU'RE WORKING SO HARD.
BUT...
ガラ…
...YOU MUST NEVER OPEN IT!
OKAY?
HUH?
UM...ER...
おろおろ
YOU CAN IF YOU WANT TO.
WHY CAN'T YOU DECIDE WHAT TO DO?
WELL, I SHOULD BE HAPPY TO TAKE THE PRESENT, BUT WHEN SHE SAID NOT TO OPEN IT...
I THOUGHT SHE WAS KIDDING.
IT APPEARS TO BE A SMALL JEWEL BOX.
SHAKE SHAKE
STOP!
YOU WILL REGRET IT!
I'LL JUST PEEK...
LISTEN TO THE PERSON WITH EXPERIENCE!

USING THE PYTHAGOREAN THEOREM TO PROVE TIME DILATION

We learned that according to the theory of relativity, time slows down for an object that is moving at a velocity close to the speed of light. But how much does time slow? When we used the Pythagorean theorem earlier, we considered this question using a triangle. Now we can consider it using a formula.

Let t denote the amount of time that has passed according to Space-Station Man looking at the rocket's light clock and t' denote the amount of time that has passed according to Rocket Man looking at his own clock (see Figure 2-1).

t: Time observed by Space-Station Man

t': Time observed by Rocket Man

Velocity of rocket relative to space station
Rocket Man
Path of light on rocket as viewed by space station
Path of light on space station as viewed by space station
Space-Station Man

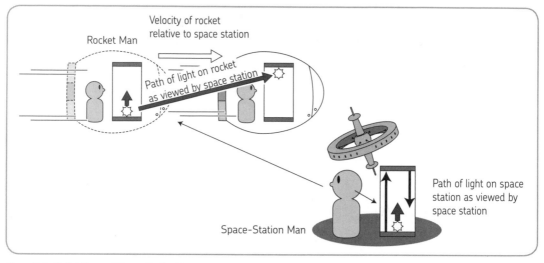

Figure 2-1: Rocket Man and Space-Station Man

When Rocket Man observes his own light clock, the light just goes up and down because the light clock is moving together with Rocket Man. Therefore, if c denotes the speed of light, when the light advances by the height of the light clock, it will have moved a distance of ct'.

Now if Space-Station Man observes the movement of the light in the rocket's light clock, the light, of course, moves at the speed of light c along an upward slanted path accompanying the movement of the rocket. That slanted line points towards the mirror (at the top) of the rocket's light clock. Measured using Space-Station Man's time t, that distance is ct. Similarly, since Space-Station Man sees the bottom of the rocket's light clock (from where the light was emitted) moving horizontally at the rocket's velocity v, the bottom will move by a distance of vt to the right in the time t that the light takes to reach the top.

This determines the three sides of a triangle (see Figure 2-2).

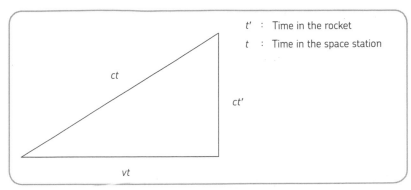

t' : Time in the rocket

t : Time in the space station

ct

ct'

vt

Figure 2-2: Distances moved by light as viewed on rocket and space station, expressed as sides of a right triangle

Therefore, from the Pythagorean theorem, we have $c^2t^2 = c^2t'^2 + v^2t^2$. Move the v^2 term to the left side of the equation:

$$\left(c^2 - v^2\right)t^2 = c^2t'^2$$

And switch the left and right sides:

$$c^2t'^2 = \left(c^2 - v^2\right)t^2$$

Dividing by c^2, we now have this:

$$t'^2 = \left(1 - \frac{v^2}{c^2}\right)t^2$$

Now take the square root of both sides and use the positive solution:

$$t' = \sqrt{1 - \frac{v^2}{c^2}} \times t$$

This is the relationship between Rocket Man's time t' and Space-Station Man's time t.

Note that $t' < t$ since $\sqrt{1 - \frac{v^2}{c^2}} < 1$.

A second (1 s) measured on the Rocket Man's light clock thus corresponds to a longer time measured by the Space-Station Man. So Space-Station Man sees the Rocket Man's clock tick off seconds at a slower rate than his own. In other words, time advances more slowly for Rocket Man than for Space-Station Man.

It is also apparent by considering the term

$$\sqrt{1 - \frac{v^2}{c^2}}$$

that this time-slowing effect is greater the closer v is to c.

With this formula, we can calculate the time dilation effect for an object moving at any relative speed.

HOW MUCH DOES TIME SLOW DOWN?

We have learned that time slows down for a moving object. Let's now determine exactly how much it slows down. We will use space travel as an example for this calculation.

As we saw above, the slowing of time is related to the velocity of the moving object. Recall that the closer the velocity of the moving object is to the speed of light, the greater the time-slowing effect becomes.

As a destination for our space travel, let's pick the star that is closest to our Sun— Alpha Centauri (see Figure 2-3). If we chose a destination inside our solar system, we wouldn't put the theory of relativity to the test because we can go to Mars or Venus in just a few years even with current technology, going nowhere near the speed of light.

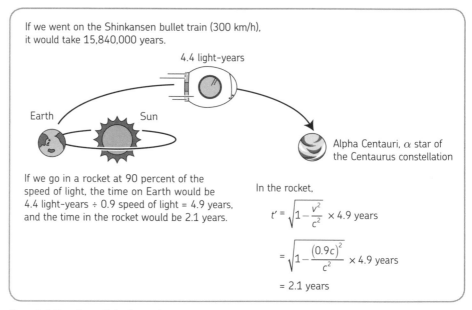

If we went on the Shinkansen bullet train (300 km/h), it would take 15,840,000 years.

4.4 light-years

Earth   Sun

Alpha Centauri, α star of the Centaurus constellation

If we go in a rocket at 90 percent of the speed of light, the time on Earth would be 4.4 light-years ÷ 0.9 speed of light = 4.9 years, and the time in the rocket would be 2.1 years.

In the rocket,

$$t' = \sqrt{1 - \frac{v^2}{c^2}} \times 4.9 \text{ years}$$

$$= \sqrt{1 - \frac{(0.9c)^2}{c^2}} \times 4.9 \text{ years}$$

$$= 2.1 \text{ years}$$

Figure 2-3: Traveling to Alpha Centauri

Alpha Centauri (the α star of the Centaurus constellation) is 4.4 light-years away from Earth. A light-year, which is the distance that light travels in one year, is approximately 9,460,800,000,000 km.[*] Traveling 4.4 light-years on the Shinkansen bullet train (at 300 km/h) would take approximately 15,840,000 years. But if we fly that distance to Alpha Centauri at 90 percent of the speed of light, the journey will take us only 2.1 years. This is despite the fact that 4.9 years will pass on Earth!

If astronauts were sent out from Earth to Alpha Centauri, even if they returned as soon as they got there, news reports on Earth would follow them for approximately 10 years, but when their families welcomed them home, they would have aged only 4.2 years.

This relativistic time slowing effect is even greater if you move faster.

To understand this relationship, let's consider traveling from the Milky Way galaxy, which contains our Sun, to the Andromeda galaxy (M31), which is near the Milky Way. The Andromeda galaxy is visible in a dark, clear winter sky as a faint smudge in the Andromeda constellation. It is approximately 2,500,000 light-years away. Because light from the galaxy takes 2,500,000 years to reach us, the Andromeda galaxy that we see now is actually the galaxy 2,500,000 years ago. Even if an explosion occurred in the Andromeda Galaxy today, we wouldn't see it until 2,500,000 years later; the light would still take that long to reach us.

If astronauts travel to the Andromeda galaxy at 99.999999999 percent of the speed of light, 11.2 years will pass for the one-way trip in the spaceship, but nearly 2,500,000 years will pass on Earth! Therefore, when the spaceship returns, although the astronauts will have aged 22.4 years, the people on Earth who will greet them will be from a time 5,000,000 years later.

Let's do the math: If a spaceship traveled to the Andromeda galaxy at 99.999999999 percent of the speed of light, people on Earth will have to wait 2,500,000 years for the spaceship to reach Andromeda and another 2,500,000 years for the spaceship to return. In all, 5 million years will pass before the spaceship returns to Earth. The amount of time that passes for the astronauts on the spaceship during this journey t' can be calculated using the time dilation formula that we derived above, using the time that it takes for the spaceship to travel to Andromeda as measured on earth ($t = 2,500,000$ years) and the velocity of the spaceship ($v = 0.99999999999c$).

$$t' = \sqrt{1 - \frac{v^2}{c^2}} \times t$$

$$t' = \sqrt{1 - \left(\frac{0.99999999999c}{c}\right)^2} \times t$$

$$t' = 11.2 \text{ years}$$

---

[*] Let's do the math: $\dfrac{300,000 \text{ km}}{1 \text{ second}} \times \dfrac{60 \text{ seconds}}{1 \text{ minute}} \times \dfrac{60 \text{ minutes}}{1 \text{ hour}} \times \dfrac{24 \text{ hours}}{1 \text{ day}} \times \dfrac{365 \text{ days}}{1 \text{ year}}$

$= 9,460,800,000 \text{ km}!$

For the astronauts on the spaceship, the round-trip to Andromeda will take 22.4 years, as shown in Figure 2-4. For the people on Earth, the astronauts will return only 22.4 years older than when they left 5 million years earlier!

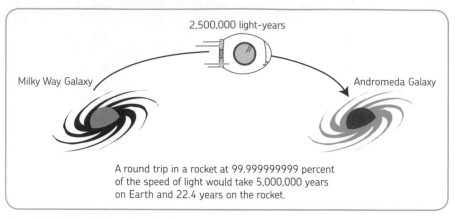

2,500,000 light-years

Milky Way Galaxy

Andromeda Galaxy

A round trip in a rocket at 99.999999999 percent of the speed of light would take 5,000,000 years on Earth and 22.4 years on the rocket.

Figure 2-4: Traveling to the Andromeda galaxy

THE FASTER AN OBJECT MOVES, THE SHORTER AND HEAVIER IT BECOMES?

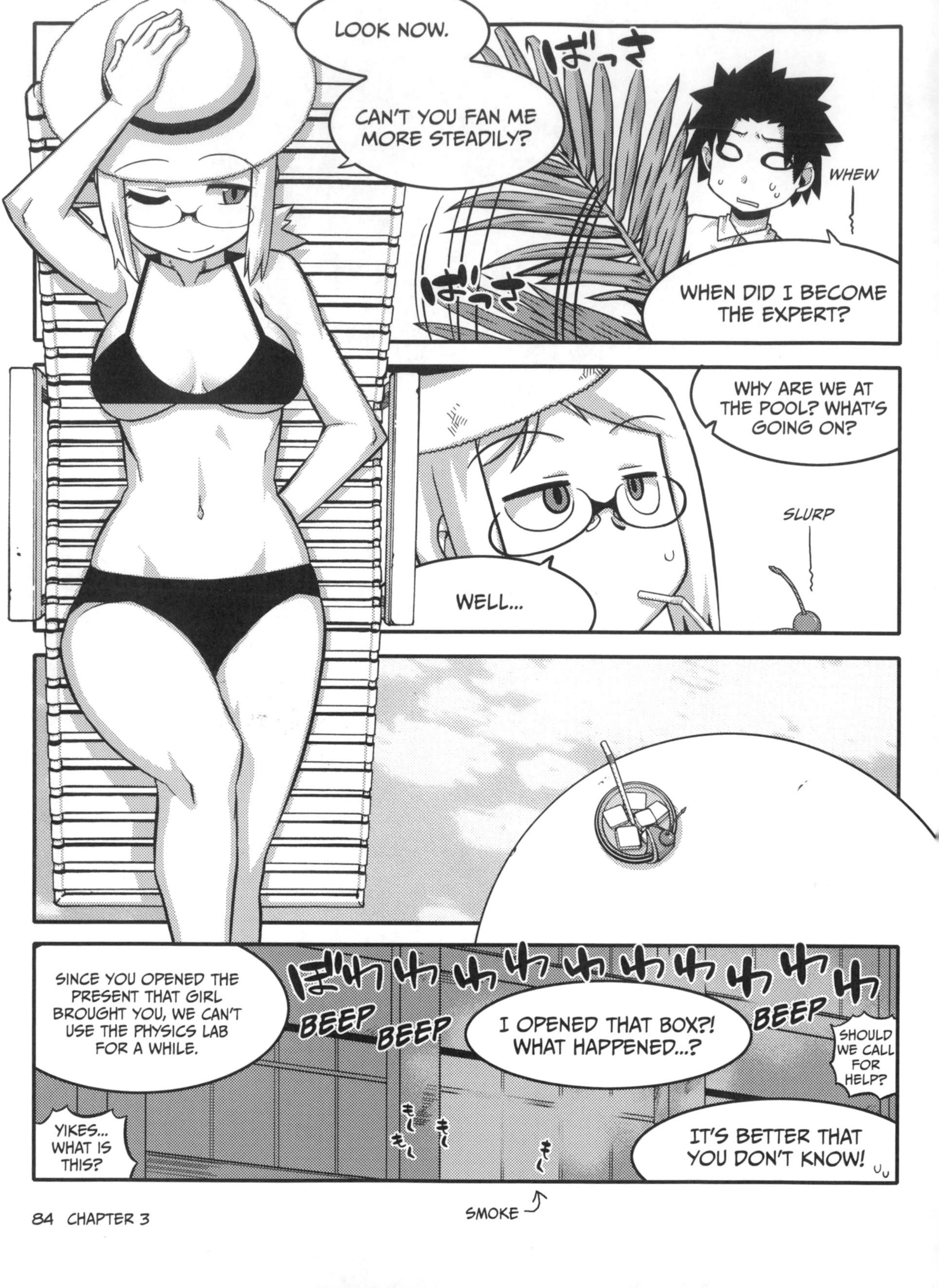
LOOK NOW.
CAN'T YOU FAN ME MORE STEADILY?
ばっさ
WHEW
ばっさ
WHEN DID I BECOME THE EXPERT?
WHY ARE WE AT THE POOL? WHAT'S GOING ON?
WELL...
SLURP
SINCE YOU OPENED THE PRESENT THAT GIRL BROUGHT YOU, WE CAN'T USE THE PHYSICS LAB FOR A WHILE.
ぼわわわわわわわわわわ
BEEP
BEEP
I OPENED THAT BOX?! WHAT HAPPENED...?
BEEP
SHOULD WE CALL FOR HELP?
YIKES... WHAT IS THIS?
IT'S BETTER THAT YOU DON'T KNOW!
SMOKE

THEREFORE, THE POOL WILL BE OUR CLASSROOM TODAY.

YOU'RE ACTUALLY PROBABLY PRETTY HAPPY ABOUT THAT, AREN'T YOU?

AM I?

YOU CAN LOOK AT YOUR BEAUTIFUL TEACHER IN HER BATHING SUIT.

THAT'S RIDICULOUS!

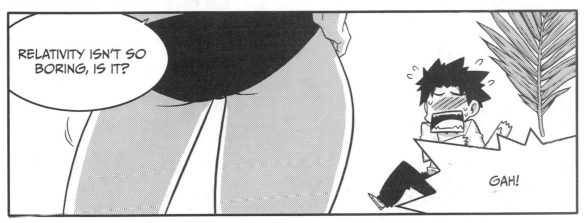

RELATIVITY ISN'T SO BORING, IS IT?

GAH!

GET ON WITH THE LESSON ALREADY, OKAY?

ALL RIGHT, ALL RIGHT.

CLINK

IN SPECIAL RELATIVITY, TIME SLOWS DOWN FOR A MOVING OBJECT, LENGTH CONTRACTS, AND MASS INCREASES, RIGHT?

SINCE WE'VE ALREADY TALKED ABOUT TIME DILATION, TODAY LET'S TALK ABOUT LENGTH AND WEIGHT.

THE FASTER AN OBJECT MOVES, THE SHORTER AND HEAVIER IT BECOMES? 85

1. DOES LENGTH CONTRACT WHEN YOU GO FASTER?

THE BOTTOM LINE IS THAT AN OBJECT MOVING AT NEAR-LIGHT SPEED IS OBSERVED TO CONTRACT IN THE DIRECTION OF MOTION.

THAT MEANS AN OBJECT CONTRACTS BECAUSE OF THE RESISTANCE OF AIR OR SOMETHING, DOESN'T IT?

SQUEAK SQUEAK

NO! THAT'S NOT WHAT HAPPENS!

SPACE *ITSELF* CONTRACTS, AND AS A RESULT, SPACE ITSELF IS OBSERVED TO CONTRACT BY A PERSON AT REST.

I SORT OF UNDERSTAND AND SORT OF DON'T UNDERSTAND...

TO MAKE IT EASIER TO UNDERSTAND, I'LL TELL YOU A STORY. COME HERE.

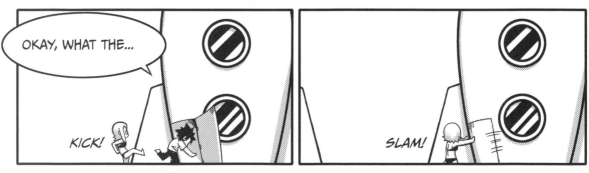

OKAY, WHAT THE...

KICK!

SLAM!

LET'S SAY THERE'S A BOMB ON BOARD...

SET TO EXPLODE IN 10 SECONDS.

LET ME OUT!!!

HOWEVER! DISASTER CAN BE AVERTED IF YOU ARRIVE AT MY SPACE STATION, 4,000,000 KM AWAY FROM THE ROCKET.

ISN'T THAT GREAT, MINAGI?

THIS IS NOT GOOD AT ALL! OR SHOULD I SAY, THIS IS ABSOLUTELY A SETUP!

WHY DOES YOUR VOICE SOUND SO HAPPY!

IT'S OKAY, DON'T WORRY... THIS ROCKET FLIES AT 270,000 KM/S.

AH, WHAT'S THAT? IF I'M GOING SO QUICKLY, I MAY GET THERE IN TIME...

RELIEVED

DISTANCE ROCKET TRAVELS IN 10 SECONDS 270,000 KM/S × 10 S = 2,700,000 KM

4,000,000 KM

OH NO!

I WON'T GET THERE IN TIME...

GOODBYE...

IS THAT REALLY TRUE?

IF YOU'RE GOING AT NEAR-LIGHT SPEED AT 270,000 KM/S, THE FLOW OF TIME RELATIVE TO WHERE I AM AT REST CHANGES, DOESN'T IT?

SNIFF

OH. THAT'S RIGHT.

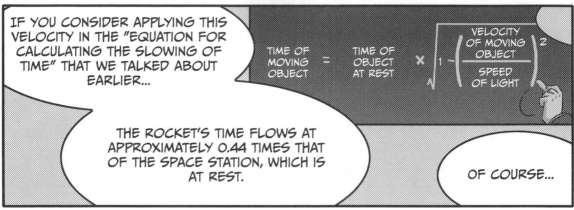

IF YOU CONSIDER APPLYING THIS VELOCITY IN THE "EQUATION FOR CALCULATING THE SLOWING OF TIME" THAT WE TALKED ABOUT EARLIER...

$$\text{TIME OF MOVING OBJECT} = \text{TIME OF OBJECT AT REST} \times \sqrt{1 - \left(\frac{\text{VELOCITY OF MOVING OBJECT}}{\text{SPEED OF LIGHT}}\right)^2}$$

THE ROCKET'S TIME FLOWS AT APPROXIMATELY 0.44 TIMES THAT OF THE SPACE STATION, WHICH IS AT REST.

OF COURSE...

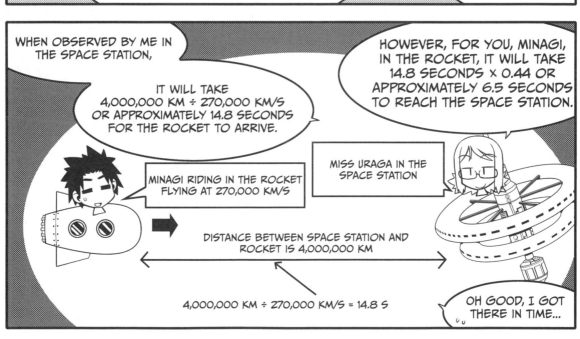

WHEN OBSERVED BY ME IN THE SPACE STATION,

IT WILL TAKE 4,000,000 KM ÷ 270,000 KM/S OR APPROXIMATELY 14.8 SECONDS FOR THE ROCKET TO ARRIVE.

HOWEVER, FOR YOU, MINAGI, IN THE ROCKET, IT WILL TAKE 14.8 SECONDS × 0.44 OR APPROXIMATELY 6.5 SECONDS TO REACH THE SPACE STATION.

MINAGI RIDING IN THE ROCKET FLYING AT 270,000 KM/S

MISS URAGA IN THE SPACE STATION

DISTANCE BETWEEN SPACE STATION AND ROCKET IS 4,000,000 KM

4,000,000 KM ÷ 270,000 KM/S = 14.8 S

OH GOOD, I GOT THERE IN TIME...

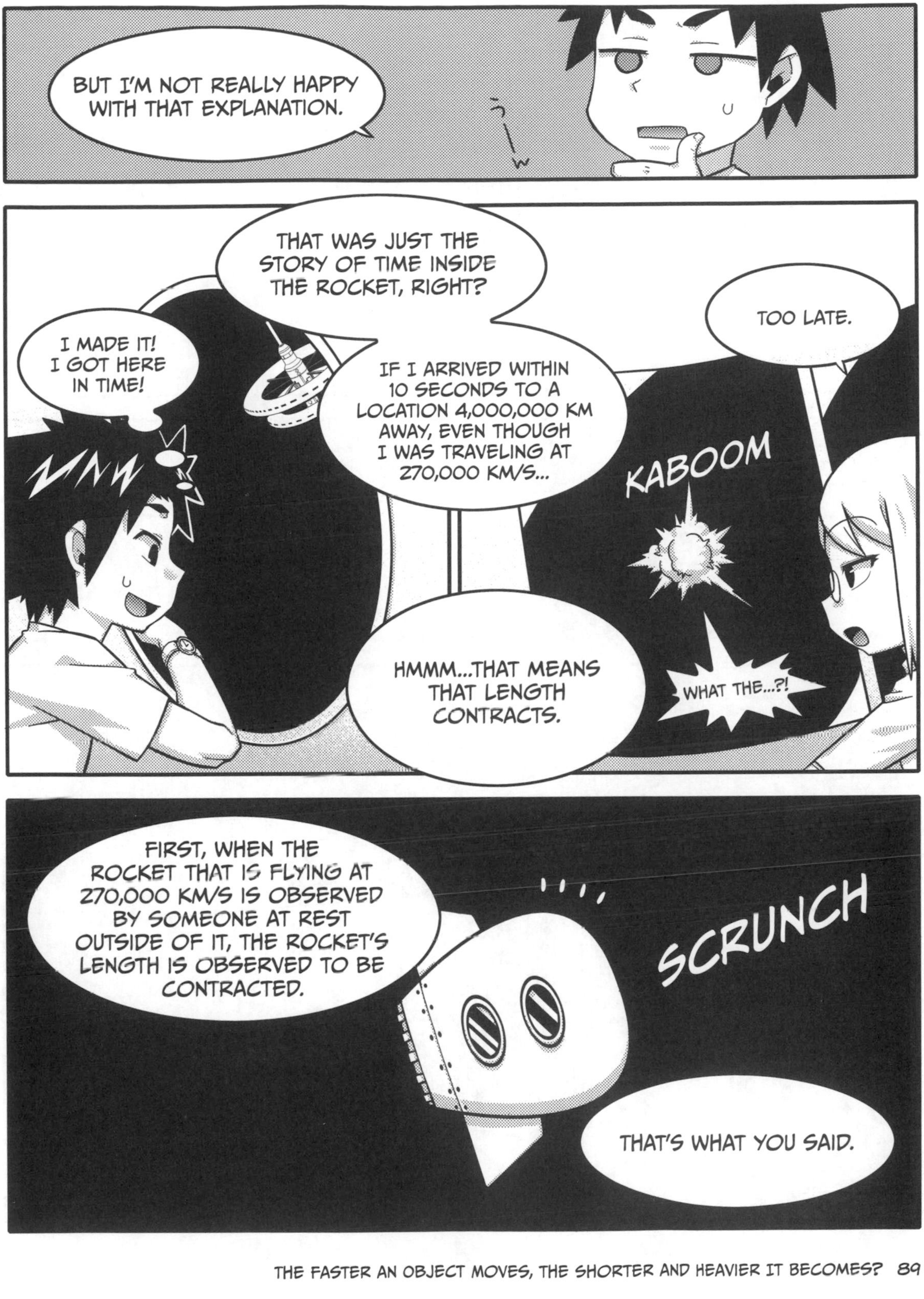

BUT I'M NOT REALLY HAPPY WITH THAT EXPLANATION.
I MADE IT! I GOT HERE IN TIME!
THAT WAS JUST THE STORY OF TIME INSIDE THE ROCKET, RIGHT?
IF I ARRIVED WITHIN 10 SECONDS TO A LOCATION 4,000,000 KM AWAY, EVEN THOUGH I WAS TRAVELING AT 270,000 KM/S...
TOO LATE.
KABOOM
WHAT THE...?!
HMMM...THAT MEANS THAT LENGTH CONTRACTS.
FIRST, WHEN THE ROCKET THAT IS FLYING AT 270,000 KM/S IS OBSERVED BY SOMEONE AT REST OUTSIDE OF IT, THE ROCKET'S LENGTH IS OBSERVED TO BE CONTRACTED.
SCRUNCH
THAT'S WHAT YOU SAID.

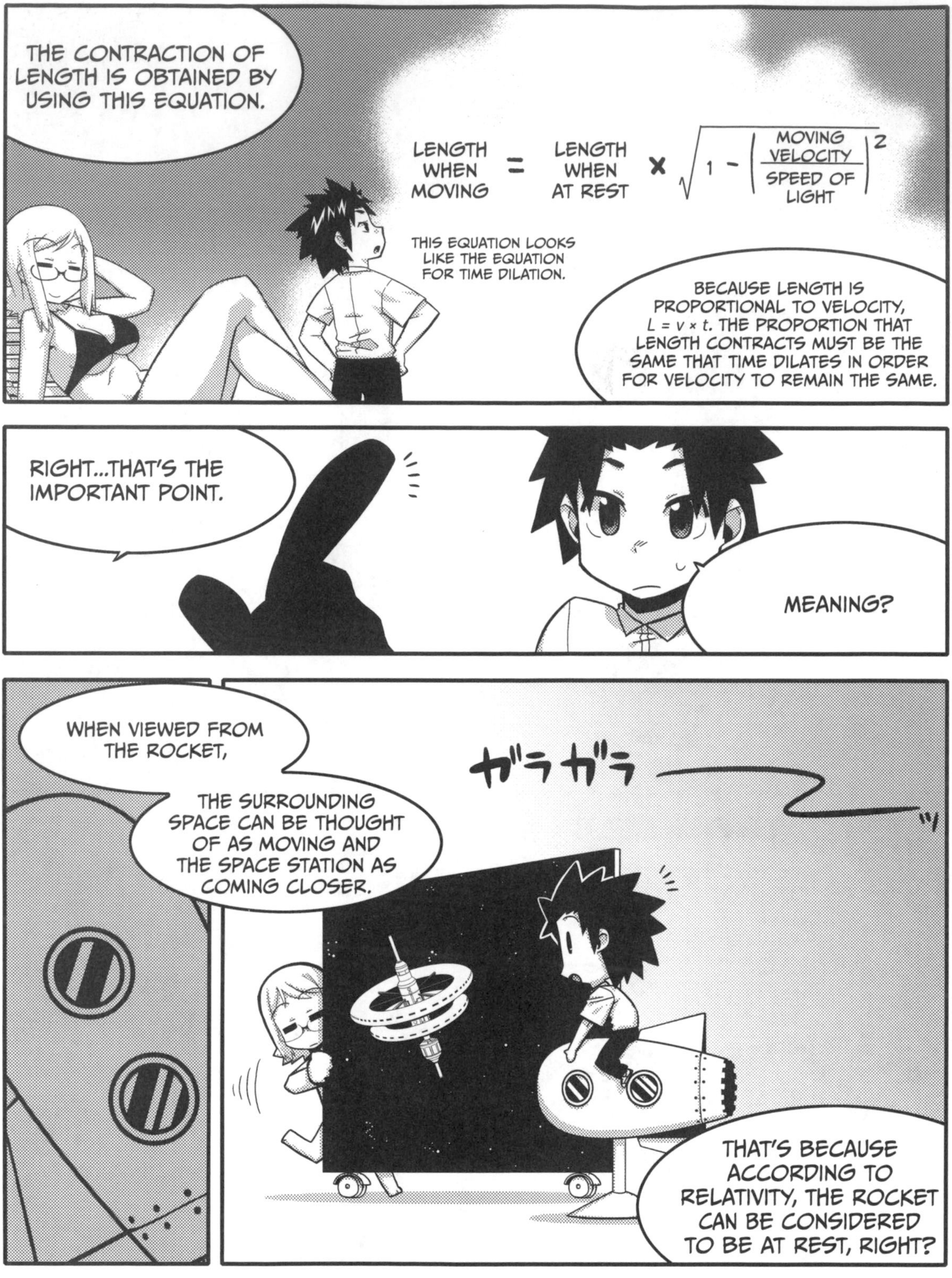
THE CONTRACTION OF LENGTH IS OBTAINED BY USING THIS EQUATION.
LENGTH WHEN MOVING = LENGTH WHEN AT REST × $\sqrt{1 - \left(\frac{\text{MOVING VELOCITY}}{\text{SPEED OF LIGHT}}\right)^2}$
THIS EQUATION LOOKS LIKE THE EQUATION FOR TIME DILATION.
BECAUSE LENGTH IS PROPORTIONAL TO VELOCITY, $L = v \times t$. THE PROPORTION THAT LENGTH CONTRACTS MUST BE THE SAME THAT TIME DILATES IN ORDER FOR VELOCITY TO REMAIN THE SAME.
RIGHT...THAT'S THE IMPORTANT POINT.
MEANING?
WHEN VIEWED FROM THE ROCKET,
THE SURROUNDING SPACE CAN BE THOUGHT OF AS MOVING AND THE SPACE STATION AS COMING CLOSER.
ガラガラ
ッ
THAT'S BECAUSE ACCORDING TO RELATIVITY, THE ROCKET CAN BE CONSIDERED TO BE AT REST, RIGHT?

THE SPACE SURROUNDING THE ROCKET CONTRACTED!

WHAT?!

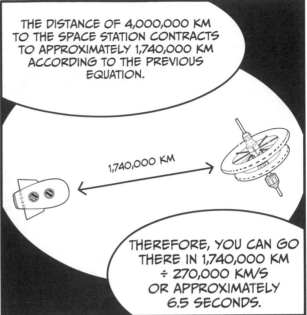

THE DISTANCE OF 4,000,000 KM TO THE SPACE STATION CONTRACTS TO APPROXIMATELY 1,740,000 KM ACCORDING TO THE PREVIOUS EQUATION.

1,740,000 KM

THEREFORE, YOU CAN GO THERE IN 1,740,000 KM ÷ 270,000 KM/S OR APPROXIMATELY 6.5 SECONDS.

IN THE THEORY OF RELATIVITY, THE CONTRACTION OF LENGTH IS JUST AS IMPORTANT AS THE SLOWING OF TIME.

I SEE...

AND TIME AND SPACE, WHICH NORMALLY ARE CONSIDERED SEPARATELY, MUST BE CONSIDERED TOGETHER.

TIME

TIME

SPACE

SPACE

THIS IS A REVOLUTIONARY PART OF RELATIVITY THEORY.

THE FASTER AN OBJECT MOVES, THE SHORTER AND HEAVIER IT BECOMES?

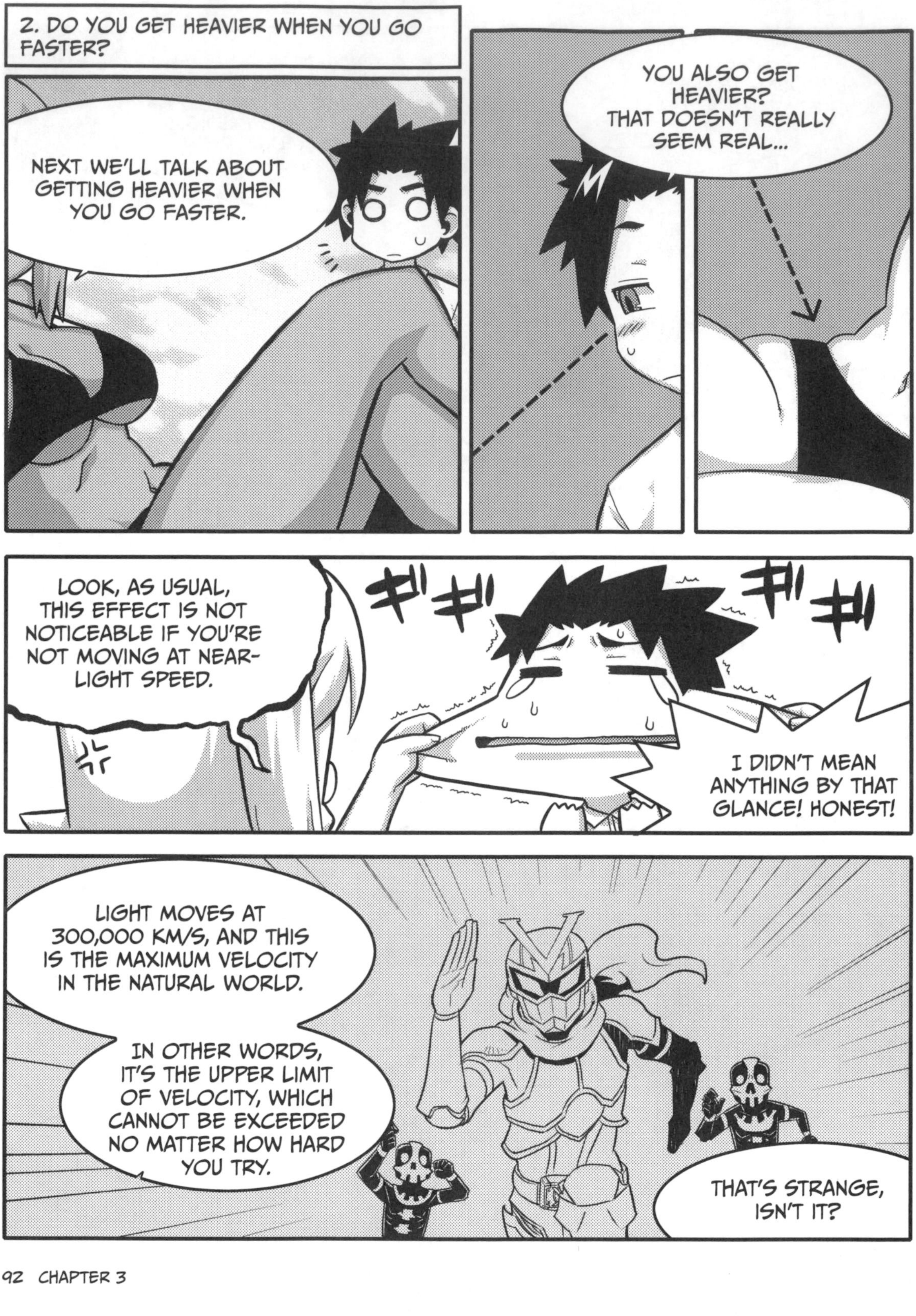
2. DO YOU GET HEAVIER WHEN YOU GO FASTER?
NEXT WE'LL TALK ABOUT GETTING HEAVIER WHEN YOU GO FASTER.
YOU ALSO GET HEAVIER? THAT DOESN'T REALLY SEEM REAL...
LOOK, AS USUAL, THIS EFFECT IS NOT NOTICEABLE IF YOU'RE NOT MOVING AT NEAR-LIGHT SPEED.
I DIDN'T MEAN ANYTHING BY THAT GLANCE! HONEST!
LIGHT MOVES AT 300,000 KM/S, AND THIS IS THE MAXIMUM VELOCITY IN THE NATURAL WORLD.
IN OTHER WORDS, IT'S THE UPPER LIMIT OF VELOCITY, WHICH CANNOT BE EXCEEDED NO MATTER HOW HARD YOU TRY.
THAT'S STRANGE, ISN'T IT?

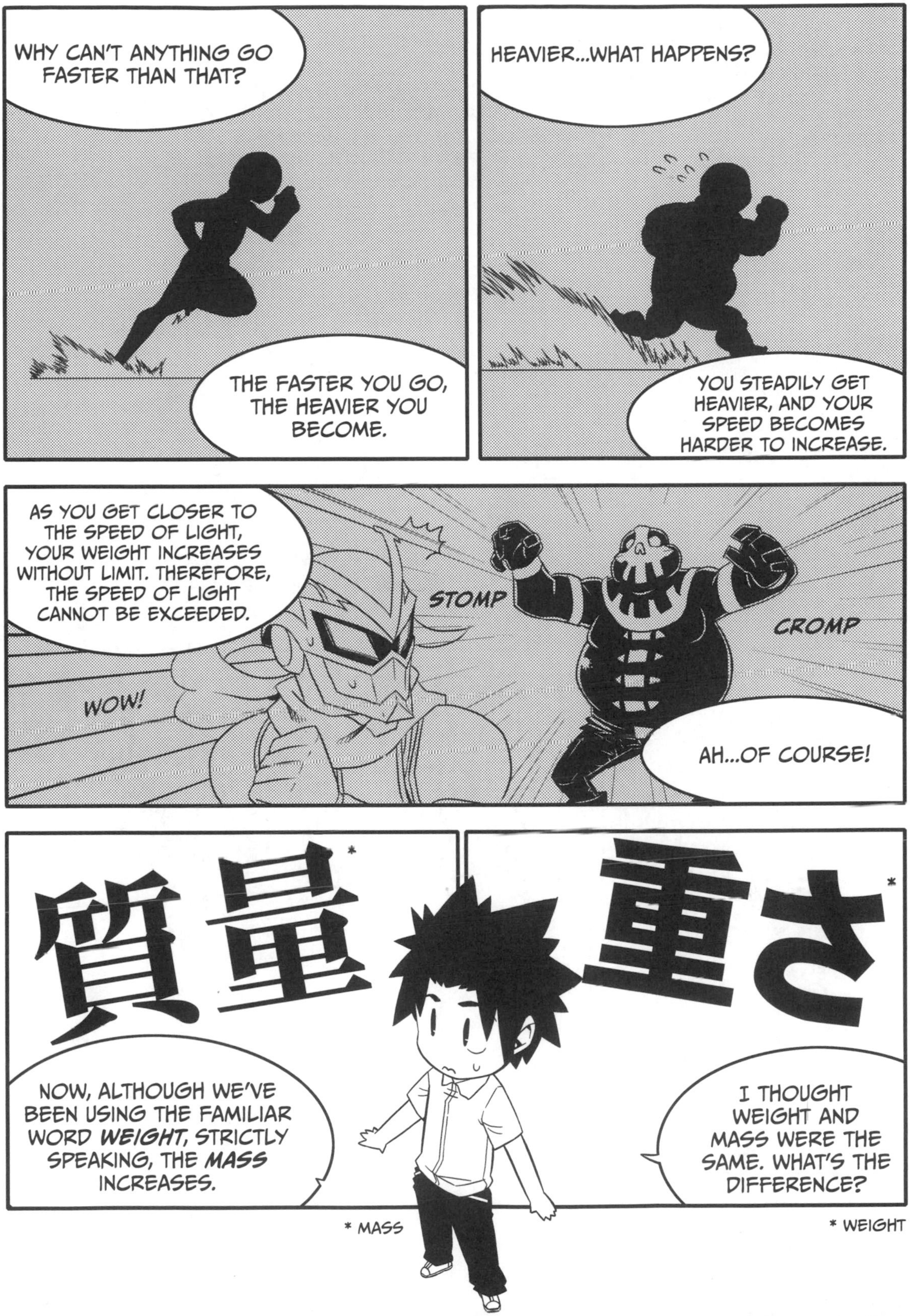
WHY CAN'T ANYTHING GO FASTER THAN THAT?
THE FASTER YOU GO, THE HEAVIER YOU BECOME.
HEAVIER...WHAT HAPPENS?
YOU STEADILY GET HEAVIER, AND YOUR SPEED BECOMES HARDER TO INCREASE.
AS YOU GET CLOSER TO THE SPEED OF LIGHT, YOUR WEIGHT INCREASES WITHOUT LIMIT. THEREFORE, THE SPEED OF LIGHT CANNOT BE EXCEEDED.
STOMP
CROMP
WOW!
AH...OF COURSE!
質量*
NOW, ALTHOUGH WE'VE BEEN USING THE FAMILIAR WORD *WEIGHT*, STRICTLY SPEAKING, THE *MASS* INCREASES.
重さ*
I THOUGHT WEIGHT AND MASS WERE THE SAME. WHAT'S THE DIFFERENCE?
* MASS
* WEIGHT

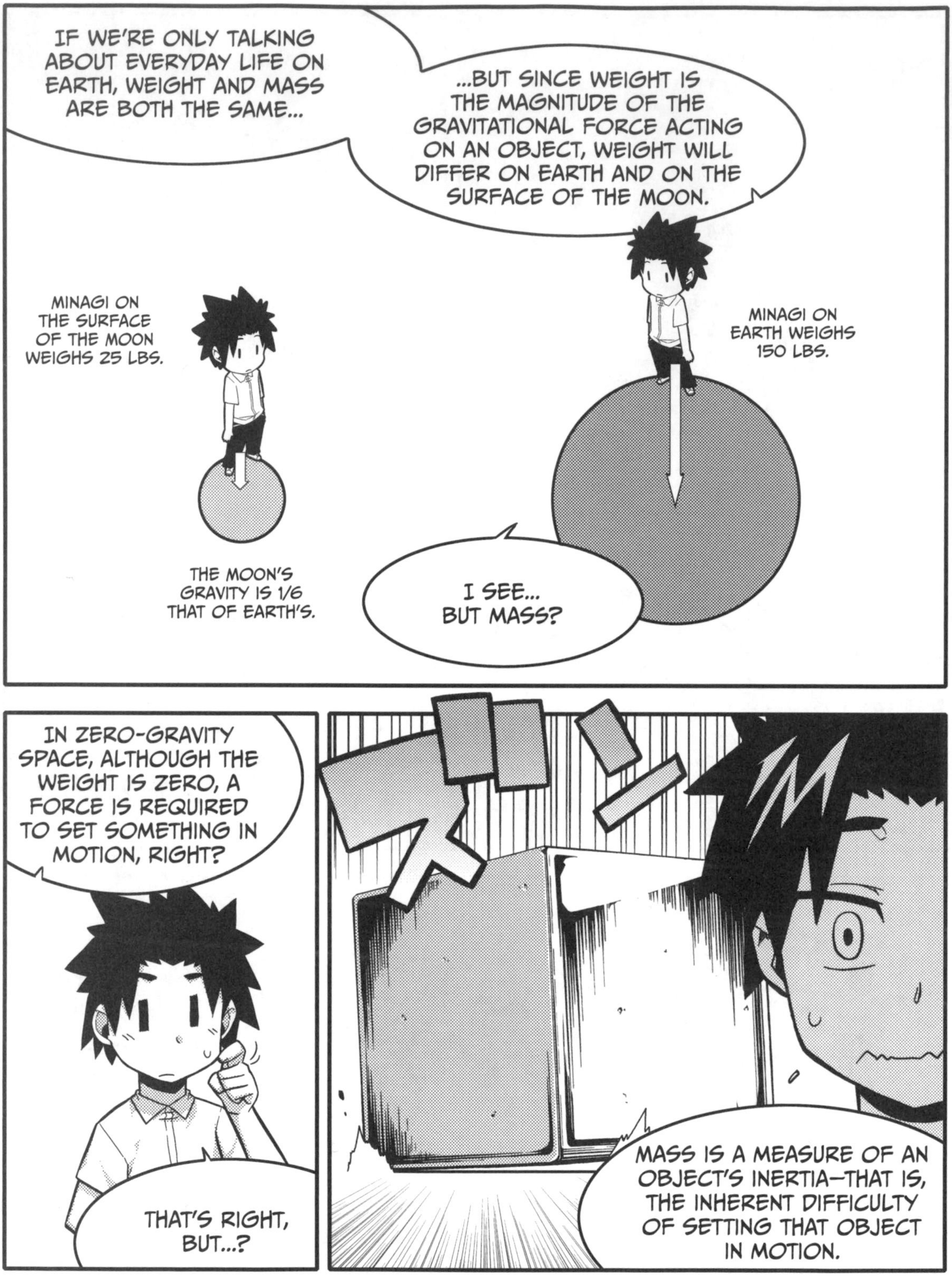
IF WE'RE ONLY TALKING ABOUT EVERYDAY LIFE ON EARTH, WEIGHT AND MASS ARE BOTH THE SAME...
...BUT SINCE WEIGHT IS THE MAGNITUDE OF THE GRAVITATIONAL FORCE ACTING ON AN OBJECT, WEIGHT WILL DIFFER ON EARTH AND ON THE SURFACE OF THE MOON.
MINAGI ON THE SURFACE OF THE MOON WEIGHS 25 LBS.
MINAGI ON EARTH WEIGHS 150 LBS.
THE MOON'S GRAVITY IS 1/6 THAT OF EARTH'S.
I SEE... BUT MASS?
IN ZERO-GRAVITY SPACE, ALTHOUGH THE WEIGHT IS ZERO, A FORCE IS REQUIRED TO SET SOMETHING IN MOTION, RIGHT?
THAT'S RIGHT, BUT...?
MASS IS A MEASURE OF AN OBJECT'S INERTIA—THAT IS, THE INHERENT DIFFICULTY OF SETTING THAT OBJECT IN MOTION.

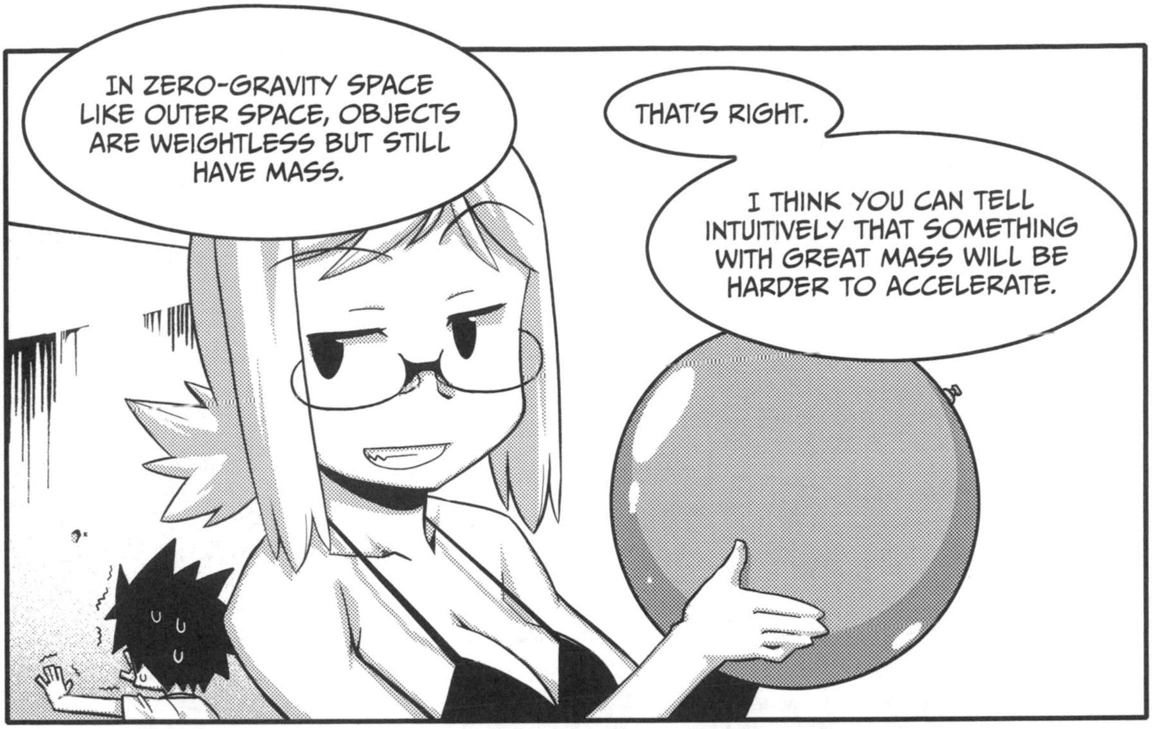
IN ZERO-GRAVITY SPACE LIKE OUTER SPACE, OBJECTS ARE WEIGHTLESS BUT STILL HAVE MASS.
THAT'S RIGHT.
I THINK YOU CAN TELL INTUITIVELY THAT SOMETHING WITH GREAT MASS WILL BE HARDER TO ACCELERATE.

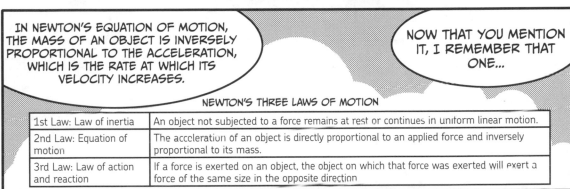

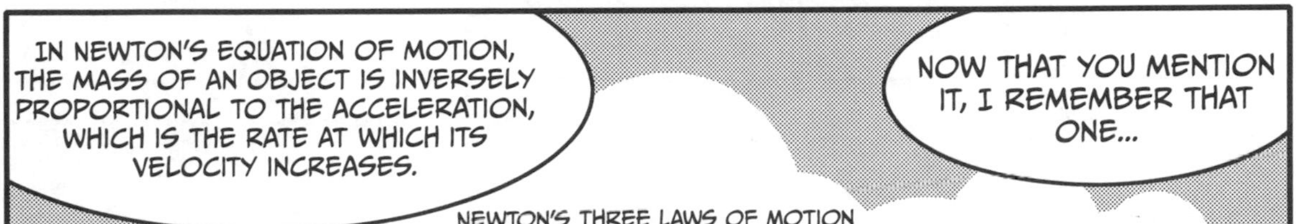
IN NEWTON'S EQUATION OF MOTION, THE MASS OF AN OBJECT IS INVERSELY PROPORTIONAL TO THE ACCELERATION, WHICH IS THE RATE AT WHICH ITS VELOCITY INCREASES.
NOW THAT YOU MENTION IT, I REMEMBER THAT ONE...

NEWTON'S THREE LAWS OF MOTION

1st Law: Law of inertia	An object not subjected to a force remains at rest or continues in uniform linear motion.
2nd Law: Equation of motion	The acceleration of an object is directly proportional to an applied force and inversely proportional to its mass.
3rd Law: Law of action and reaction	If a force is exerted on an object, the object on which that force was exerted will exert a force of the same size in the opposite direction

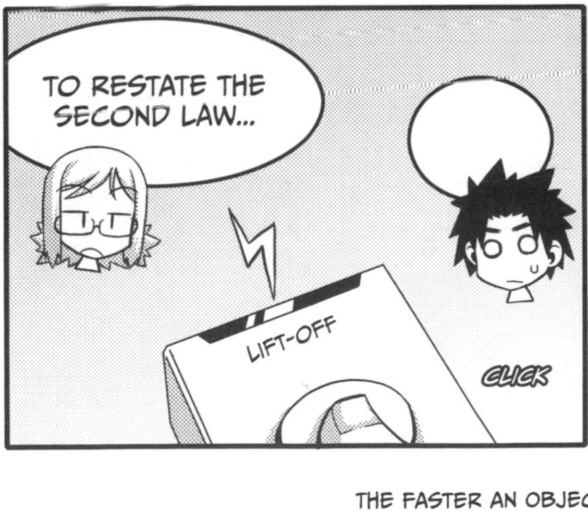
TO RESTATE THE SECOND LAW...
LIFT-OFF
CLICK

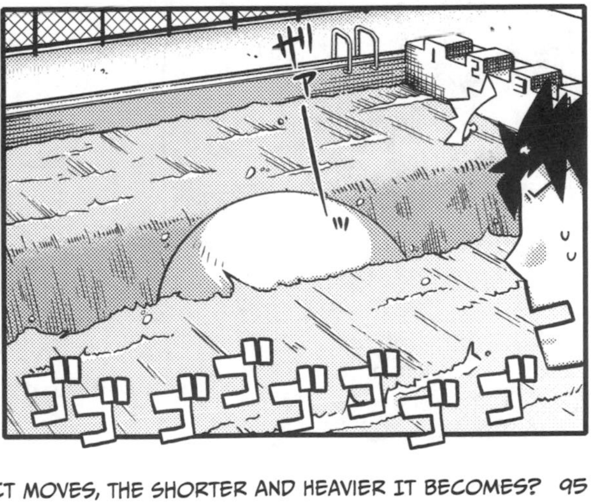

IF THE PROPULSIVE FORCE DOES NOT CHANGE FOR A ROCKET FLYING WITH AN ACCELERATION OF 10 M/S²

BUT IF ITS MASS DOUBLES, THE ACCELERATION WILL BECOME 5 M/S².

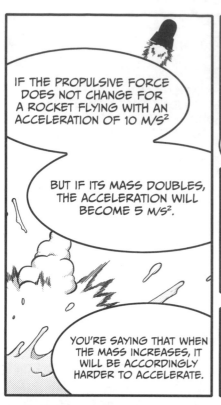

YOU'RE SAYING THAT WHEN THE MASS INCREASES, IT WILL BE ACCORDINGLY HARDER TO ACCELERATE.

THE CLOSER THE ROCKET GETS TO THE SPEED OF LIGHT, THE GREATER ITS MASS BECOMES...

AND IF YOU TRY TO FURTHER ACCELERATE THE ROCKET BY EXERTING ADDITIONAL FORCE...

UNGH

...THE MASS WILL ALSO INCREASE FURTHER, AND IT WILL BECOME MORE AND MORE DIFFICULT TO ACCELERATE.

IF YOU CONTINUE IN THIS WAY, THE MASS WILL INCREASE WITHOUT LIMIT AS THE ROCKET'S VELOCITY GETS CLOSER TO THE SPEED OF LIGHT.

RELATIONSHIP BETWEEN VELOCITY AND MASS

MASS

100000000.00
10000000.00
1000000.00
100000.00
10000.00
1000.00
100.00
10.00
1.00

0.8800 0.9000 0.9200 0.9400 0.9600 0.9800 1.0000
VELOCITY RATIO (VELOCITY/SPEED OF LIGHT)

IF WE ASSUME FOR ARGUMENT'S SAKE THAT THE ROCKET'S VELOCITY REACHED THE SPEED OF LIGHT...

ITS MASS WOULD BECOME INFINITELY LARGE. THEREFORE, THE SPEED OF LIGHT CANNOT BE EXCEEDED.

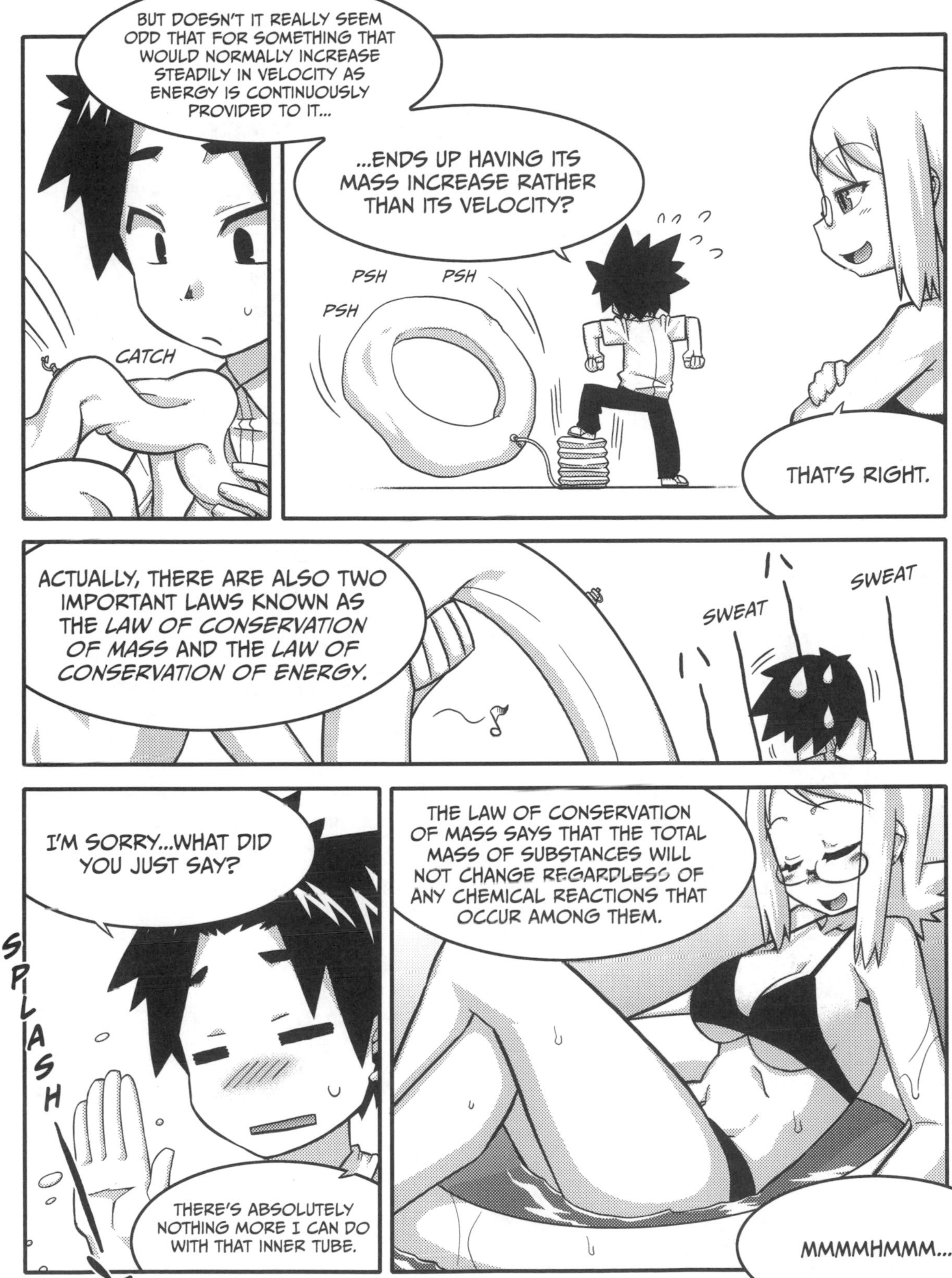
BUT DOESN'T IT REALLY SEEM ODD THAT FOR SOMETHING THAT WOULD NORMALLY INCREASE STEADILY IN VELOCITY AS ENERGY IS CONTINUOUSLY PROVIDED TO IT...
CATCH
...ENDS UP HAVING ITS MASS INCREASE RATHER THAN ITS VELOCITY?
PSH
PSH
PSH
THAT'S RIGHT.
ACTUALLY, THERE ARE ALSO TWO IMPORTANT LAWS KNOWN AS THE *LAW OF CONSERVATION OF MASS* AND THE *LAW OF CONSERVATION OF ENERGY.*
SWEAT
SWEAT
I'M SORRY...WHAT DID YOU JUST SAY?
SPLASH
THERE'S ABSOLUTELY NOTHING MORE I CAN DO WITH THAT INNER TUBE.
THE LAW OF CONSERVATION OF MASS SAYS THAT THE TOTAL MASS OF SUBSTANCES WILL NOT CHANGE REGARDLESS OF ANY CHEMICAL REACTIONS THAT OCCUR AMONG THEM.
MMMMHMMM...

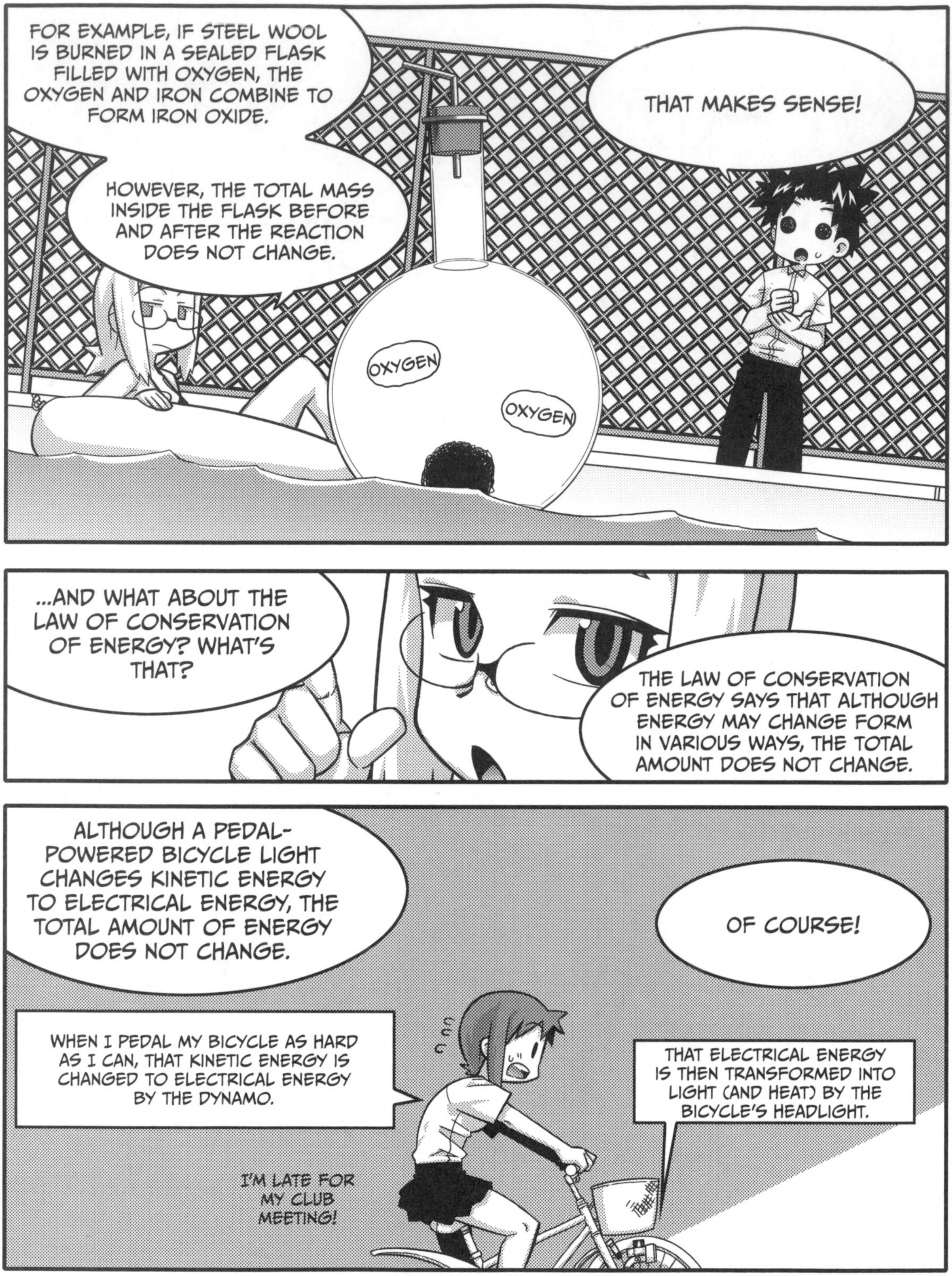
FOR EXAMPLE, IF STEEL WOOL IS BURNED IN A SEALED FLASK FILLED WITH OXYGEN, THE OXYGEN AND IRON COMBINE TO FORM IRON OXIDE.
HOWEVER, THE TOTAL MASS INSIDE THE FLASK BEFORE AND AFTER THE REACTION DOES NOT CHANGE.
OXYGEN
OXYGEN
THAT MAKES SENSE!
...AND WHAT ABOUT THE LAW OF CONSERVATION OF ENERGY? WHAT'S THAT?
THE LAW OF CONSERVATION OF ENERGY SAYS THAT ALTHOUGH ENERGY MAY CHANGE FORM IN VARIOUS WAYS, THE TOTAL AMOUNT DOES NOT CHANGE.
ALTHOUGH A PEDAL-POWERED BICYCLE LIGHT CHANGES KINETIC ENERGY TO ELECTRICAL ENERGY, THE TOTAL AMOUNT OF ENERGY DOES NOT CHANGE.
OF COURSE!
WHEN I PEDAL MY BICYCLE AS HARD AS I CAN, THAT KINETIC ENERGY IS CHANGED TO ELECTRICAL ENERGY BY THE DYNAMO.
I'M LATE FOR MY CLUB MEETING!
THAT ELECTRICAL ENERGY IS THEN TRANSFORMED INTO LIGHT (AND HEAT) BY THE BICYCLE'S HEADLIGHT.

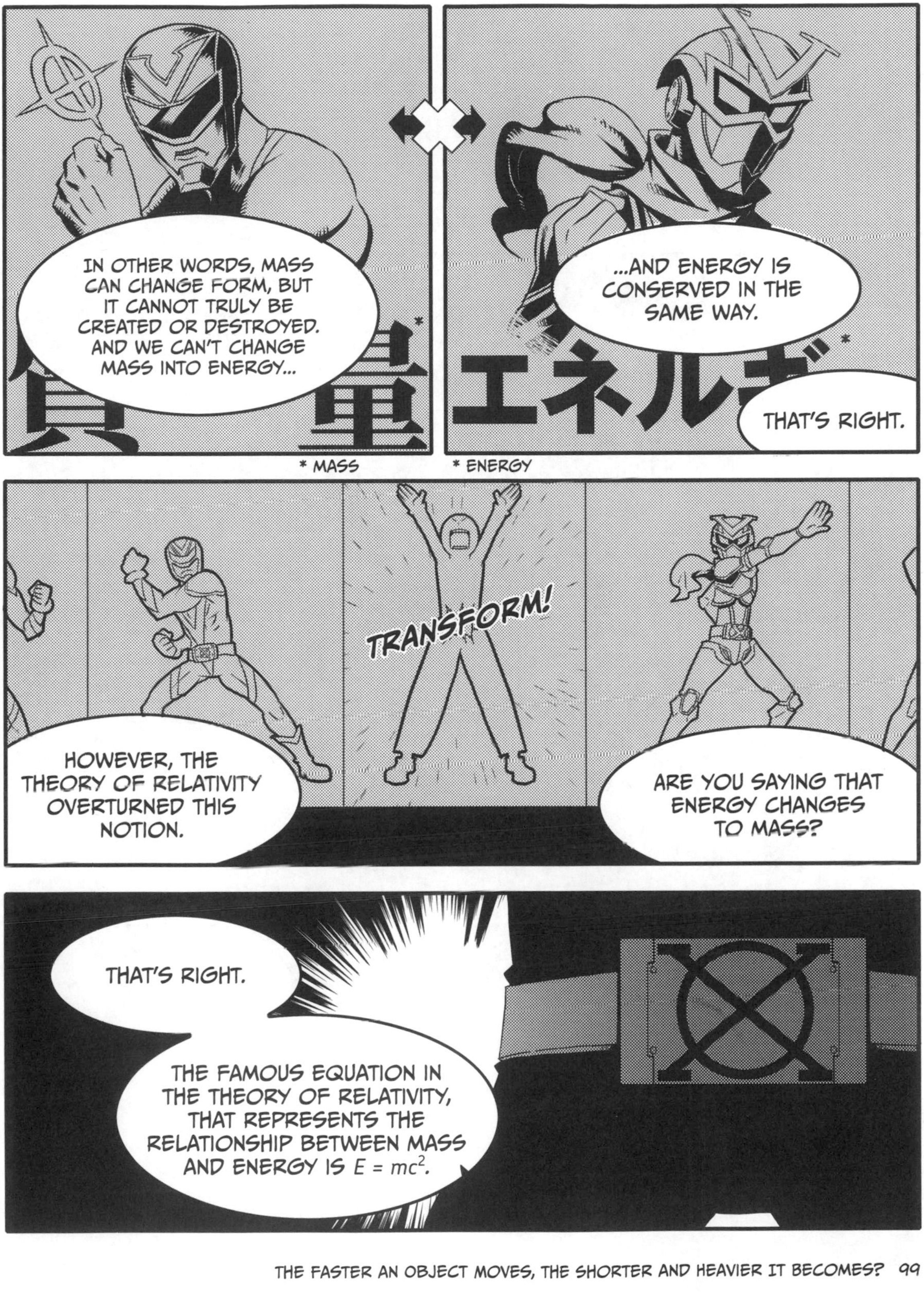
IN OTHER WORDS, MASS CAN CHANGE FORM, BUT IT CANNOT TRULY BE CREATED OR DESTROYED. AND WE CAN'T CHANGE MASS INTO ENERGY...
* MASS
...AND ENERGY IS CONSERVED IN THE SAME WAY.
THAT'S RIGHT.
* ENERGY
HOWEVER, THE THEORY OF RELATIVITY OVERTURNED THIS NOTION.
TRANSFORM!
ARE YOU SAYING THAT ENERGY CHANGES TO MASS?
THAT'S RIGHT.
THE FAMOUS EQUATION IN THE THEORY OF RELATIVITY, THAT REPRESENTS THE RELATIONSHIP BETWEEN MASS AND ENERGY IS $E = mc^2$.

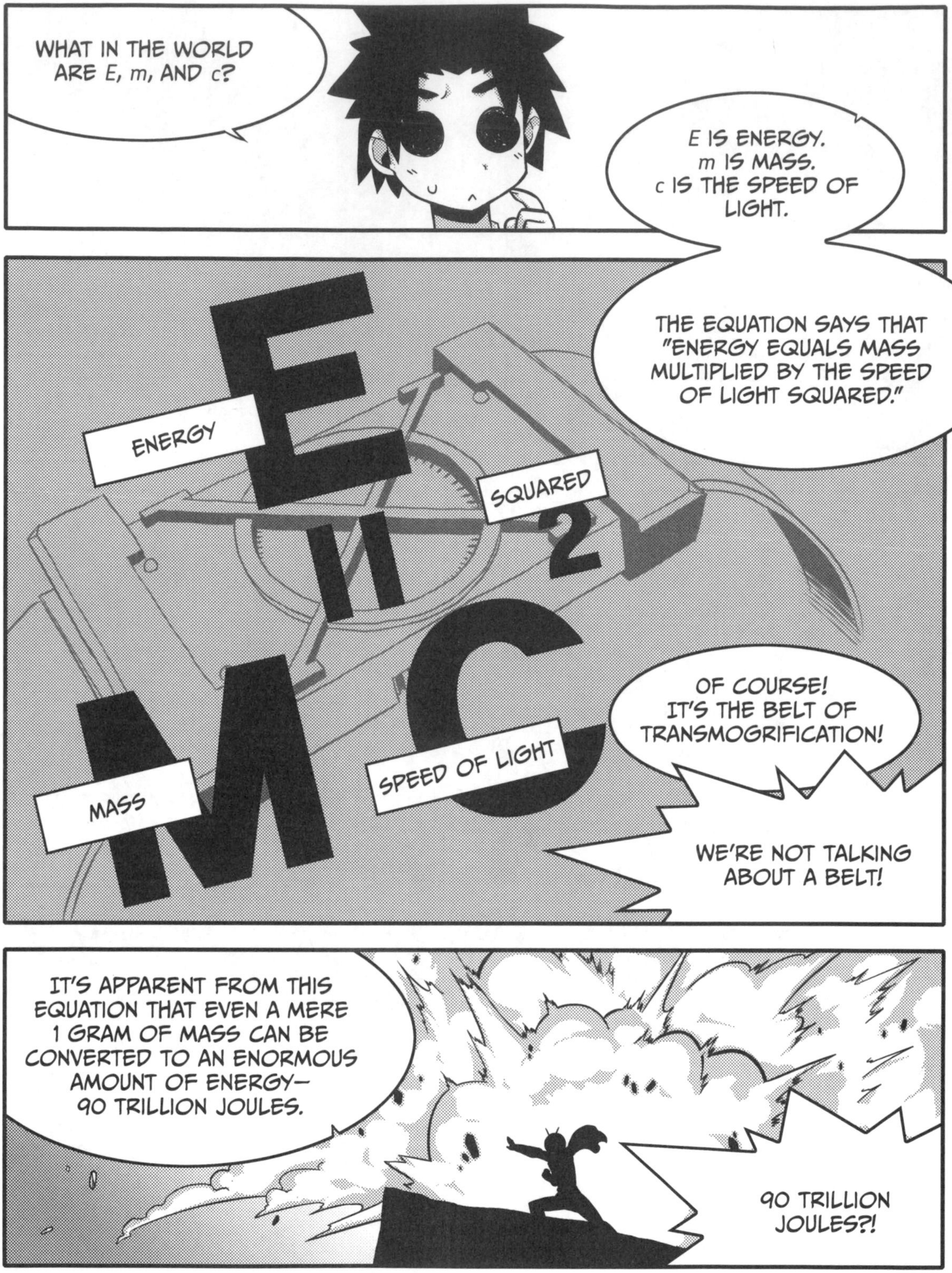
WHAT IN THE WORLD ARE *E*, *m*, AND *c*?
E IS ENERGY. *m* IS MASS. *c* IS THE SPEED OF LIGHT.
E
ENERGY
=
SQUARED
2
M
MASS
C
SPEED OF LIGHT
THE EQUATION SAYS THAT "ENERGY EQUALS MASS MULTIPLIED BY THE SPEED OF LIGHT SQUARED."
OF COURSE! IT'S THE BELT OF TRANSMOGRIFICATION!
WE'RE NOT TALKING ABOUT A BELT!
IT'S APPARENT FROM THIS EQUATION THAT EVEN A MERE 1 GRAM OF MASS CAN BE CONVERTED TO AN ENORMOUS AMOUNT OF ENERGY— 90 TRILLION JOULES.
90 TRILLION JOULES?!

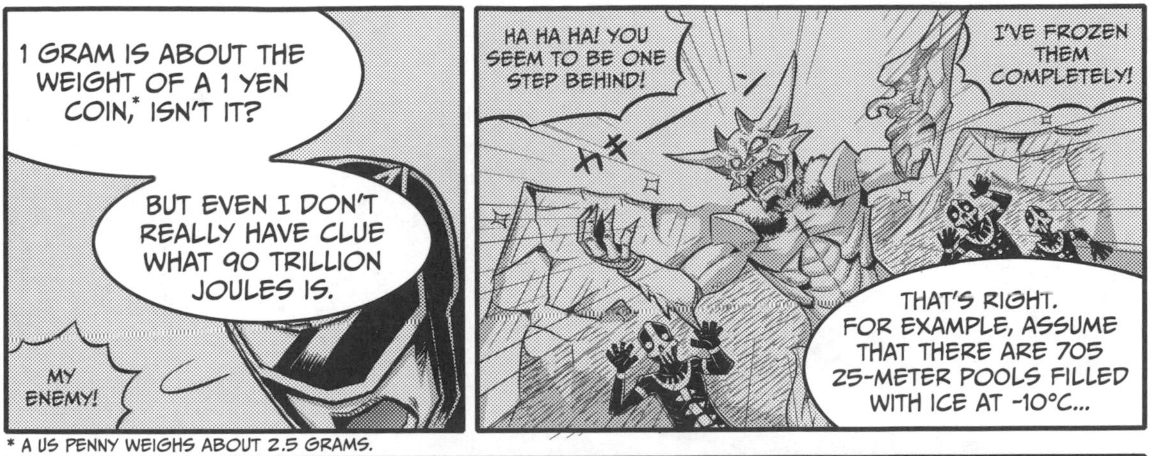
1 GRAM IS ABOUT THE WEIGHT OF A 1 YEN COIN,* ISN'T IT?
BUT EVEN I DON'T REALLY HAVE CLUE WHAT 90 TRILLION JOULES IS.
MY ENEMY!
HA HA HA! YOU SEEM TO BE ONE STEP BEHIND!
I'VE FROZEN THEM COMPLETELY!
カキ
THAT'S RIGHT. FOR EXAMPLE, ASSUME THAT THERE ARE 705 25-METER POOLS FILLED WITH ICE AT -10°C...

* A US PENNY WEIGHS ABOUT 2.5 GRAMS.

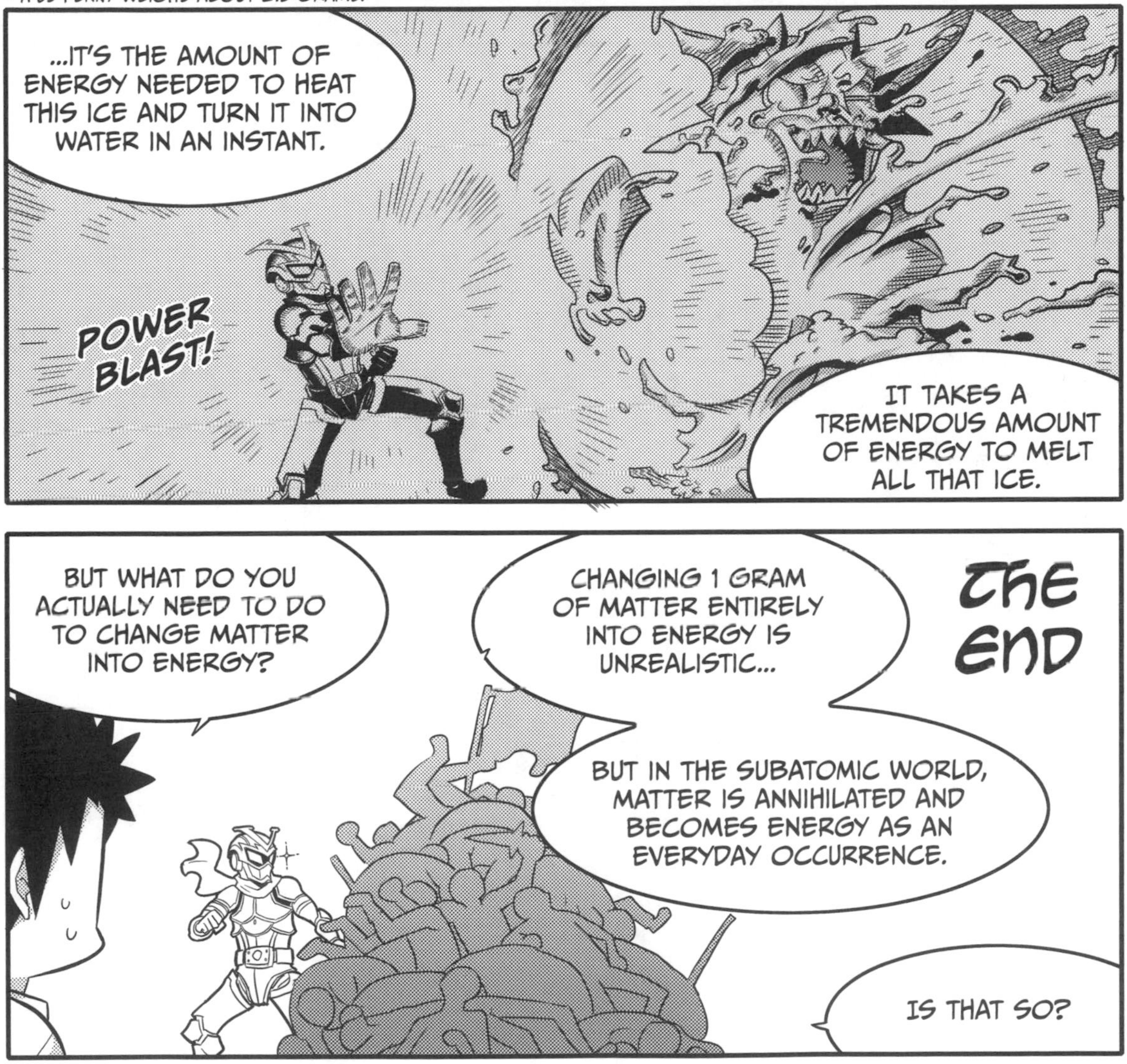
...IT'S THE AMOUNT OF ENERGY NEEDED TO HEAT THIS ICE AND TURN IT INTO WATER IN AN INSTANT.
POWER BLAST!
IT TAKES A TREMENDOUS AMOUNT OF ENERGY TO MELT ALL THAT ICE.
BUT WHAT DO YOU ACTUALLY NEED TO DO TO CHANGE MATTER INTO ENERGY?
CHANGING 1 GRAM OF MATTER ENTIRELY INTO ENERGY IS UNREALISTIC...
THE END
BUT IN THE SUBATOMIC WORLD, MATTER IS ANNIHILATED AND BECOMES ENERGY AS AN EVERYDAY OCCURRENCE.
IS THAT SO?

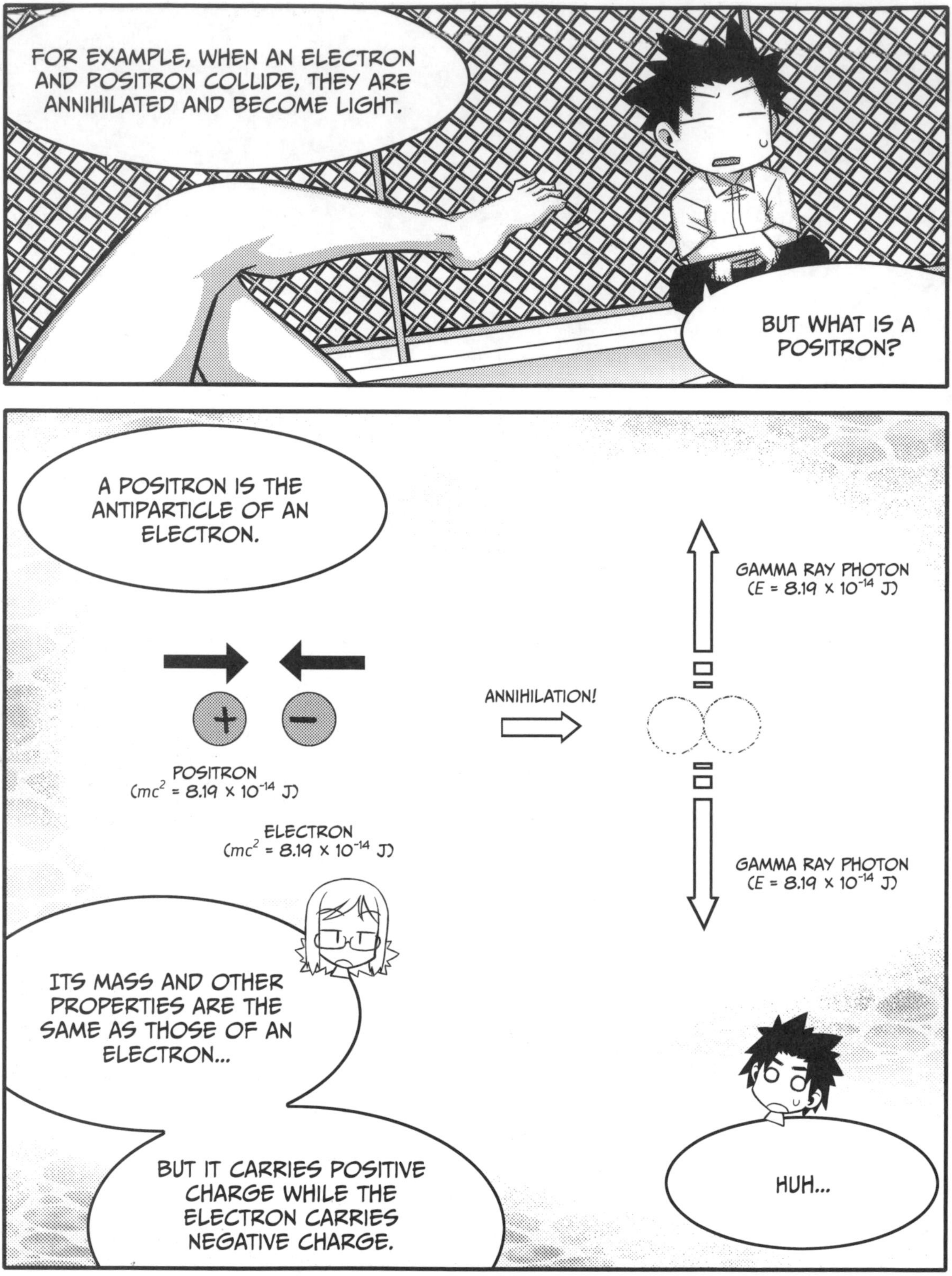
FOR EXAMPLE, WHEN AN ELECTRON AND POSITRON COLLIDE, THEY ARE ANNIHILATED AND BECOME LIGHT.
BUT WHAT IS A POSITRON?
A POSITRON IS THE ANTIPARTICLE OF AN ELECTRON.
POSITRON (mc^2 = 8.19 × 10^{-14} J)
ELECTRON (mc^2 = 8.19 × 10^{-14} J)
ANNIHILATION!
GAMMA RAY PHOTON (E = 8.19 × 10^{-14} J)
GAMMA RAY PHOTON (E = 8.19 × 10^{-14} J)
ITS MASS AND OTHER PROPERTIES ARE THE SAME AS THOSE OF AN ELECTRON...
BUT IT CARRIES POSITIVE CHARGE WHILE THE ELECTRON CARRIES NEGATIVE CHARGE.
HUH...

WHEN A PARTICLE AND ANTIPARTICLE SUCH AS AN ELECTRON AND POSITRON COLLIDE, THEY ARE ANNIHILATED.

THIS PHENOMENON IS CALLED *PAIR ANNIHILATION.*

WHEN A PARTICLE AND ANTIPARTICLE COLLIDE, THEY COMPLETELY VANISH?

ANTIPARTICLE

PARTICLE

THEY DON'T JUST "VANISH." THEY ARE CHANGED INTO THE EQUIVALENT AMOUNT OF ENERGY.

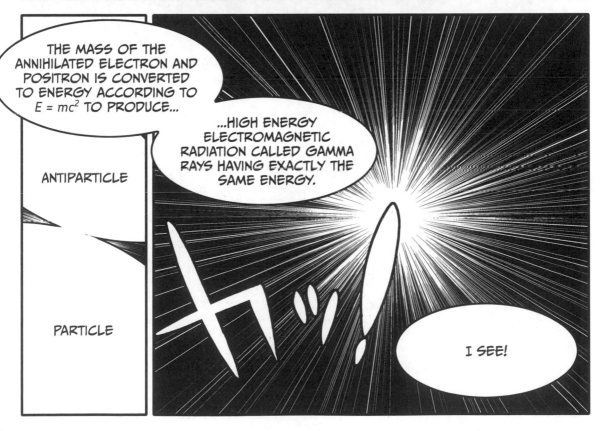

THE MASS OF THE ANNIHILATED ELECTRON AND POSITRON IS CONVERTED TO ENERGY ACCORDING TO $E = mc^2$ TO PRODUCE...

...HIGH ENERGY ELECTROMAGNETIC RADIATION CALLED GAMMA RAYS HAVING EXACTLY THE SAME ENERGY.

ANTIPARTICLE

PARTICLE

I SEE!

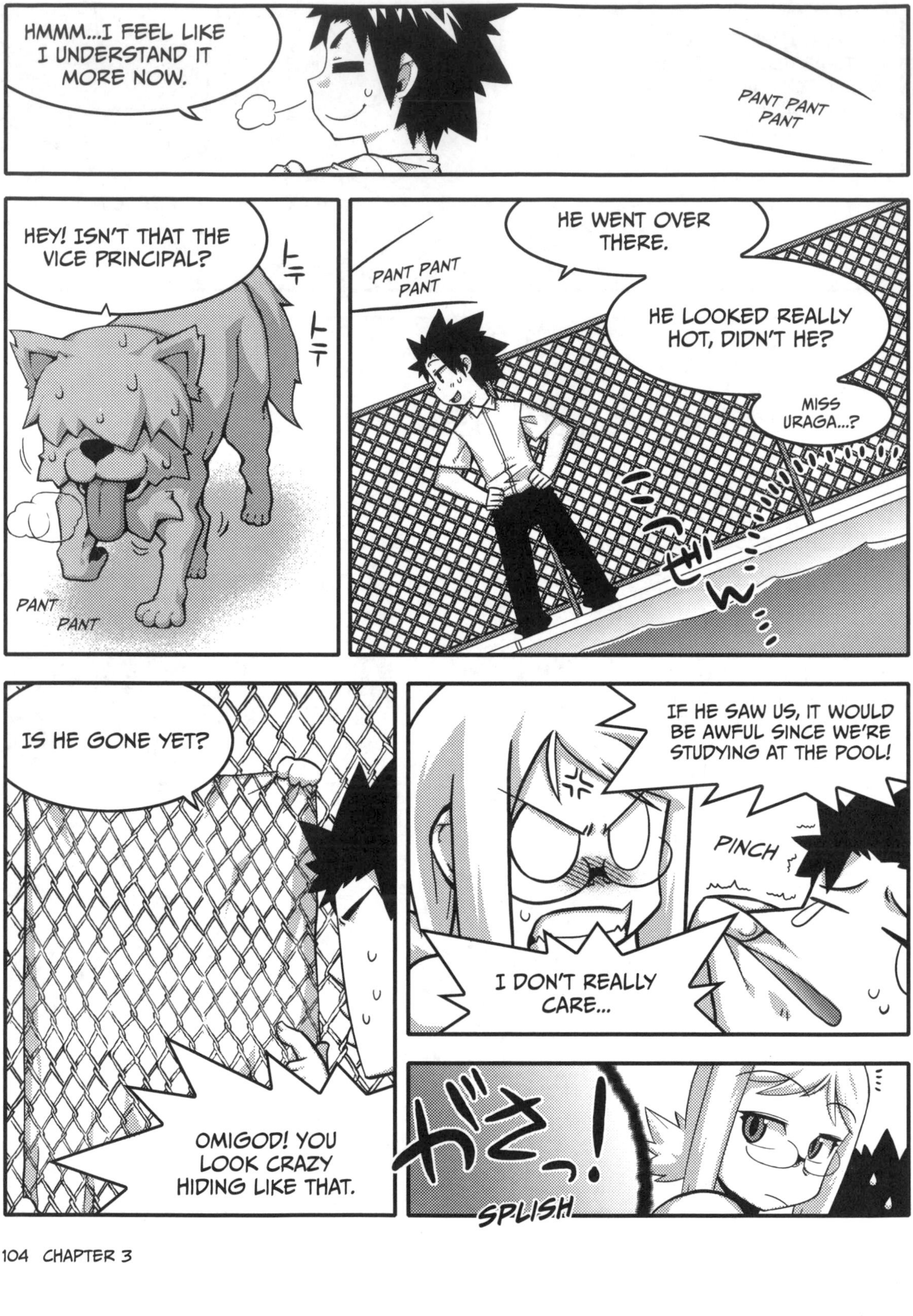
HMMM...I FEEL LIKE I UNDERSTAND IT MORE NOW.
PANT PANT PANT
HEY! ISN'T THAT THE VICE PRINCIPAL?
PANT PANT
PANT PANT PANT
HE WENT OVER THERE.
HE LOOKED REALLY HOT, DIDN'T HE?
MISS URAGA...?
IS HE GONE YET?
OMIGOD! YOU LOOK CRAZY HIDING LIKE THAT.
IF HE SAW US, IT WOULD BE AWFUL SINCE WE'RE STUDYING AT THE POOL!
PINCH
I DON'T REALLY CARE...
SPLISH

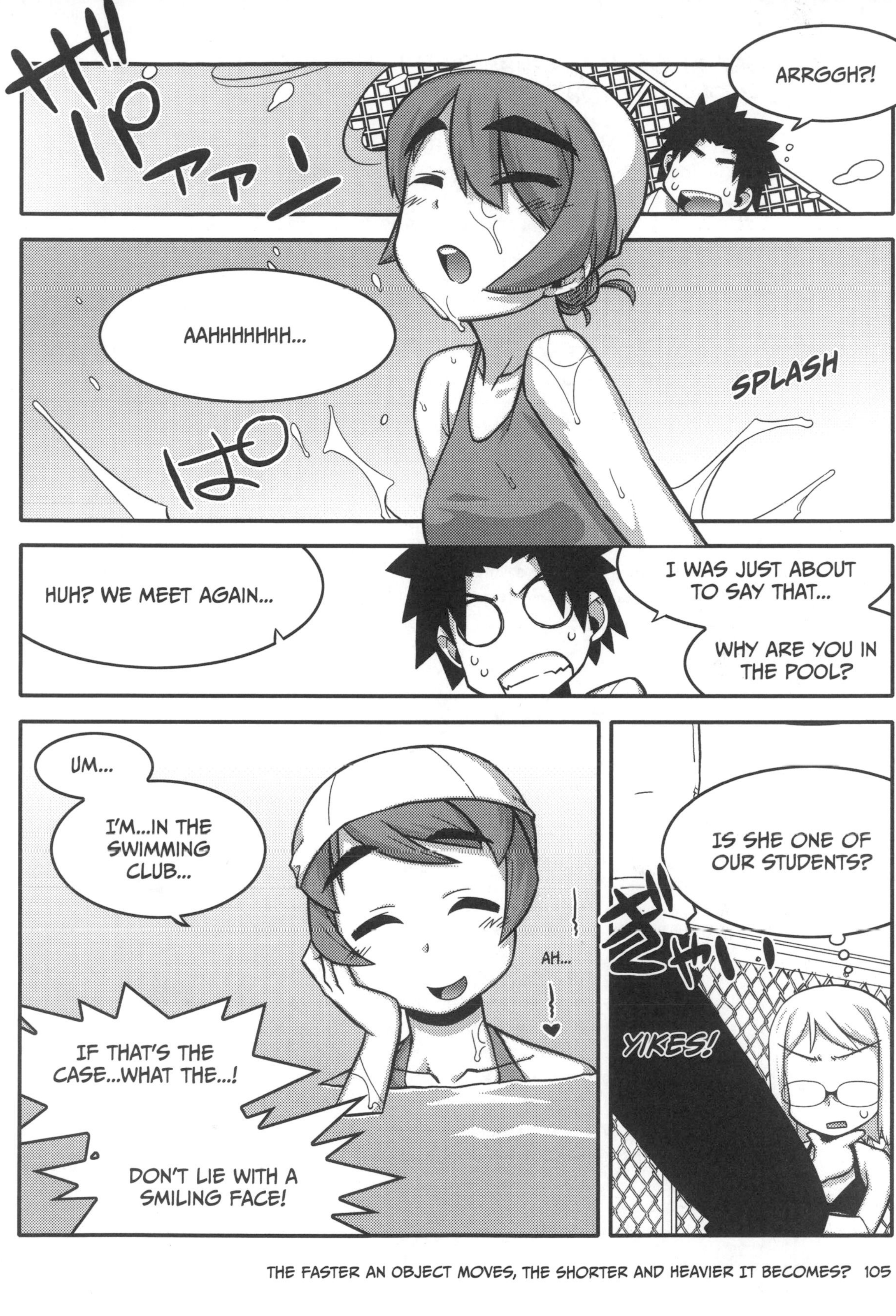
ARRGGH?!
AAHHHHHHH...
SPLASH
HUH? WE MEET AGAIN...
I WAS JUST ABOUT TO SAY THAT...
WHY ARE YOU IN THE POOL?
UM...
I'M...IN THE SWIMMING CLUB...
AH...
IF THAT'S THE CASE...WHAT THE...!
DON'T LIE WITH A SMILING FACE!
IS SHE ONE OF OUR STUDENTS?
YIKES!

USING AN EQUATION TO UNDERSTAND LENGTH CONTRACTION (LORENTZ CONTRACTION)

Let's use an equation to see how length contracts.
In this case, let's assume that a rocket is flying at a constant velocity v (see Figure 3-1).

Mr. A in the rocket thinks he's at rest.

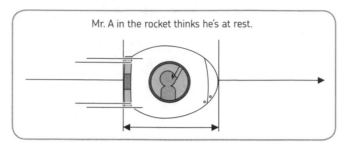

Figure 3-1: A person riding in the rocket measures the positions of the front and back ends of the rocket.

When the person riding in the rocket measures the positions of the front and back ends of the rocket, he finds the front end is at position x'_2, and the back end is at position x'_1. Therefore, the rocket's length is $l_0 = x'_2 - x'_1$.

Now what happens if this situation is observed from outside the rocket, for example, from a space station as in Figure 3-2?

Direction of motion
Velocity v
x'_1
l_0
x'_2
Mr. B in the space station

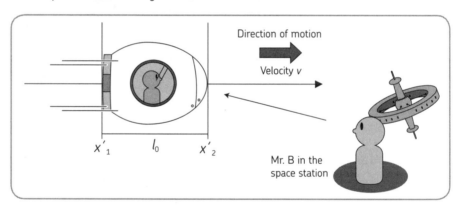

Figure 3-2: The rocket viewed from a space station

To calculate the contraction in the length of a rocket as it moves past an observer at close to the speed of light, let's consider two points in the rocket's frame of reference: x'_1 at the front of the ship and x'_2 at the back of the ship. Using the Lorentz transformation, introduced in "Wait a Second—What Happens with the Addition of Velocities?" on page 48, we can calculate how an observer who watches the ship pass measures the points at the front x_1 and back of the ship x_2, in his reference frame. The length that the observer on the outside of the ship measures will be shorter than length that the astronaut measures. This effect, *relativistic length contraction*, comes from the contraction of space at speeds close to the speed of light.

Using the Lorentz transformation $x' = \dfrac{x - vt}{\sqrt{1 - \left(\dfrac{v}{c}\right)^2}}$, we calculate the following positions:

$$x_1' = \frac{x_1 - vt_1}{\sqrt{1 - \left(\dfrac{v}{c}\right)^2}}$$

$$x_2' = \frac{x_2 - vt_2}{\sqrt{1 - \left(\dfrac{v}{c}\right)^2}}$$

If we let $l = x_2 - x_1$ represent the rocket's length as observed from outside the rocket, then since

$$l_0 = x_2' - x_1' = \frac{x_2 - vt_2}{\sqrt{1 - \left(\dfrac{v}{c}\right)^2}} - \frac{x_1 - vt_1}{\sqrt{1 - \left(\dfrac{v}{c}\right)^2}} = \frac{\left(x_2 - x_1\right) - \left(t_2 - t_1\right)v}{\sqrt{1 - \left(\dfrac{v}{c}\right)^2}}$$

is measured at the same time, $t_1 - t_2 = 0$ because $t_2 = t_1$, so l_0 is calculated as follows:

$$l_0 = \frac{\left(x_2 - x_1\right) - \left(t_2 - t_1\right)v}{\sqrt{1 - \left(\dfrac{v}{c}\right)^2}} = \frac{\left(x_2 - x_1\right)}{\sqrt{1 - \left(\dfrac{v}{c}\right)^2}} = \frac{l}{\sqrt{1 - \left(\dfrac{v}{c}\right)^2}}$$

The length of the spaceship measured from outside the spaceship,

$$l = l_0 \sqrt{1 - \left(\frac{v}{c}\right)^2}$$

is therefore less than the length l_0 of the same spaceship measured from inside the ship. We know this to be true because the coefficient

$$\sqrt{1 - \left(\frac{v}{c}\right)^2} < 1$$

due to the fact that the speed of the spaceship must be slower than the speed of light ($v < c$).

MUONS WITH EXTENDED LIFE SPANS

Our discussion of time slowing down and length contracting is not just a theoretical proposition. The slowing of time is observed every day.

High-energy elementary particles called *cosmic rays* are raining down on Earth all day, every day. When those cosmic rays collide with molecules in Earth's upper atmosphere, muons are generated with a certain probability. A *muon* is a type of elementary particle that is similar to an electron. The life span of a muon is approximately 2 millionths of a second in a laboratory on the ground at rest. Therefore, when a muon is produced in the upper atmosphere, several tens to several hundreds of kilometers from the ground, it would fly only 300,000 km/s × 2/1,000,000 s = 0.6 km, even if it were flying at a velocity extremely close to the speed of light. Based on these calculations, it should not reach the surface of Earth. But muons are observed on the surface of Earth! This seemingly impossible event occurs because the muon's life span is extended according to the special theory of relativity (see Figure 3-3). The extension of the lifetime of muons by time dilation has been verified in the laboratory by accelerating muons to near the speed of light.

Several 10s to several 100s km
Muons are produced when elementary particles (cosmic rays) at velocities near the speed of light collide with moledcules in Earth's upper atmosphere.
The life span of a muon is approximately 2/1,000,000 seconds.
Even if a muon were flying at near-light speed, it would fly only 300,000 km/s × 2/1,000,000 s = 0.6 km.
But muons are observed on the surface of Earth.
The life span of the muon is extended (time slows down in accordance with relativity).
Muon
Earth
Upper Atmosphere

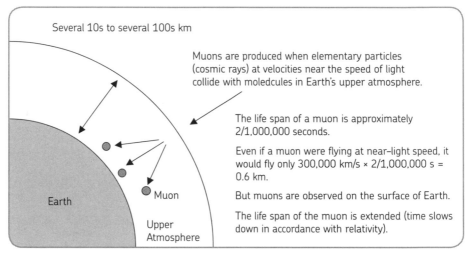

Figure 3-3: Muon life span

Let's now apply the concept of length contraction to the example of the muon. In the reference frame of the muon, the muon's lifetime is not extended—it is still 2 millionths of a second. From the reference frame of the muon, Earth is rushing toward the particle at close to the speed of light. The length between Earth and the muon contracts, however, as shown in Figure 3-4. And because the distance between Earth and the muon contracts, the muon reaches the planet's surface within its lifetime

Both the dilation of time from Earth's perspective and the contraction of length from the muon's perspective are consistent. In this way, the dilation of time and the contraction of length change together according to the theory of relativity.

Several 10s km
When viewed from the perspective of a muon at rest, the distance to the surface of Earth is several 10s km.
When viewed from the perspective of a muon traveling at near-light speed, the distance to the surface of Earth contracts to less than 0.6 km.
Contracts to 0.6 km
Earth
Upper Atmosphere
To the muon, Earth appears to be approaching at a velocity close to the speed of light.

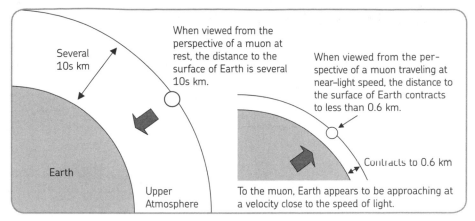

Figure 3-4: Distance contracts as well as time.

MASS WHEN MOVING

Now let's consider how the mass of an object is related to its velocity according to relativity. Let's start by reviewing the laws of motion. Before relativity was understood, the Galilean transformation and Newton's law of motion were used to describe motion.

GALILEAN TRANSFORMATION

The Galilean transformation describes the relationship between coordinate systems moving at velocity v:

$$x' = x - vt \text{ and } t' = t$$

where x' and t' represent position and time, respectively, in one system and x, v, and t represent position, velocity, and time, respectively, in the other system.

NEWTON'S SECOND LAW OF MOTION

Newton's second law of motion is represented as follows:

$$f = ma = m\frac{d^2x}{dt^2},$$

where f represents force, m represents mass, a represents acceleration, and acceleration can be considered the second derivative of displacement with respect to time:

$$a = \frac{d^2x}{dt^2}$$

According to Galileo's principle of relativity, the laws of physics operate exactly the same way, whether measured while at rest or while moving. In other words, whether you toss a ball in the air inside an elevator that is at rest or inside an elevator that is moving at a constant velocity, the ball will move up and down and return to your hand in the same way.

Now, let's look at the laws of motion in two different reference frames and verify that the laws of physics do not change when we don't take into account relativity. We'll look at the laws of motion in a reference frame that is at rest, in which the position of the ball is measured x, and in the reference frame of the moving elevator, in which the position of the ball is measured x' (see Figure 3-5).

Elevator rising at a constant velocity
x
x'
Gravity is always acting.
Tossing a ball near an elevator that is at rest at the bottom of its shaft

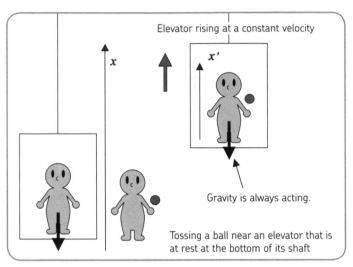

Figure 3-5: Elevator moving at a constant velocity

In this case, the velocity of the ball, which is moving in the x' direction inside the elevator, is given by

$$\frac{dx'}{dt'}$$

If we substitute the Galilean transformation $x' = x - vt$ here, we obtain the following:

$$\frac{dx'}{dt'} = \frac{d}{dt'}(x - vt) = \frac{dx}{dt'} - v\frac{dt}{dt'} = \frac{dx}{dt} - v$$

We used the relationship $\frac{dt'}{dt} = 1$ here, because $dt' = dt$.

If we differentiate again, we obtain the following:

$$\frac{d^2x'}{dt'^2} = \frac{d}{dt'}\left(\frac{dx}{dt} - v\right) = \frac{d^2x}{dt^2}$$

Since the only force acting on the ball here is gravity, if we let g denote gravity, we have this:

$$g = f = ma = m\frac{d^2x}{dt^2}$$

Note that g is a force, not accleration due to gravity in this equation. Now, if we let a' denote the acceleration inside the elevator, which is moving at a constant velocity, and let f' denote force, we have

$$m\frac{d^2x}{dt^2} = m\frac{d^2x'}{dt'^2} = ma' = f' = g$$

and the form of the equation of motion is unchanged. Since the form of the equation of motion does not change, the laws of physics remain the same.

Let's consider how the situation described above changes when we consider relativity and replace the Galilean transformation with the Lorentz transformation.

LORENTZ TRANSFORMATION

$$x' = \frac{x - vt}{\sqrt{1 - \left(\frac{v}{c}\right)^2}}$$

$$t' = \frac{t - \frac{v}{c^2}x}{\sqrt{1 - \left(\frac{v}{c}\right)^2}}$$

As we have shown in this chapter, time and space become intermixed within the framework of relativity. Therefore, when describing the coordinates of an object, it is not sufficient to give its position in three-dimensional space (x, y, z)—we must also consider its time (t). Because the units of position and time are different (meters and seconds, respectively), we multiply time by the speed of light so that we can describe the coordinate position of an object with four dimensions that all have the same unit: length. The following shows that two reference frames are mutually transformed; note that time and space are transformed together.

$$\left(ct, x, y, z\right) \longleftrightarrow \left(ct', x', y', z'\right)$$

If we use this thinking to extend the equation of motion so that its form does not change even when a Lorentz transformation is used, it is apparent that mass, which had been considered constant in Newtonian mechanics, is represented in a form similar to the Lorentz transformation:

$$m = \frac{m_0}{\sqrt{1 - \left(\dfrac{v}{c}\right)^2}}$$

Here, m_0, which is called the *rest mass* or *invariant mass*, is the mass measured in a coordinate system at rest ($v = 0$).

RELATIONSHIP BETWEEN ENERGY AND MASS

In the same way, if we consider energy in a form that matches the Lorentz transformation, it is represented as follows:

$$E = \frac{m_0 c^2}{\sqrt{1 - \left(\dfrac{v}{c}\right)^2}}$$

If we substitute the earlier relationship,

$$m = \frac{m_0}{\sqrt{1 - \left(\dfrac{v}{c}\right)^2}}$$

we derive the famous relationship between energy and mass: $E = mc^2$.

Now when $|x| \ll 1$, if we use the approximation $(1 + x)^\alpha \approx 1 + \alpha x$ under the condition

$$\left(\frac{v}{c}\right)^2 \ll 1$$

(velocity v is sufficiently small compared with the speed of light), we obtain the following:

$$E = \frac{m_0 c^2}{\sqrt{1 - \left(\dfrac{v}{c}\right)^2}} = m_0 c^2 \left[1 - \left(\frac{v}{c}\right)^2\right]^{-\frac{1}{2}} \cong m_0 c^2 \left[1 + \frac{1}{2}\left(\frac{v}{c}\right)^2\right] = m_0 c^2 + \frac{1}{2} m_0 v^2$$

The total energy of an object is the sum of its kinetic energy

$$E_k = \frac{1}{2} m_0 v^2$$

and its rest energy ($E = mc^2$). This means that even when an object is not moving, it has energy associated with its mass. Rest energy $E = mc^2$ is similar in form to the kinetic energy in Newtonian mechanics, where

$$E_k = \frac{1}{2} m_0 v^2$$

and $m_0 c^2$ is called the *rest energy.*

DOES LIGHT HAVE ZERO MASS?

The equation that we derived above for the mass of an object in motion,

$$m = \frac{m_0}{\sqrt{1 - \left(\frac{v}{c}\right)^2}}$$

tells us that as the velocity of an object approaches the speed of light, its energy approaches infinity (see Figure 3-6). Therefore, the only way that light can exist (without having infinite energy) is if its mass is 0.

100000000.00
10000000.00
1000000.00
100000.00
10000.00
1000.00
100.00
10.00
1.00
Mass (kg)
0.8800
0.9000
0.9200
0.9400
0.9600
0.9800
1.0000
Velocity / Speed of light

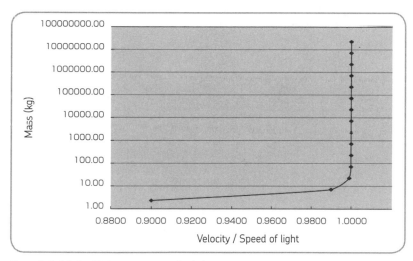

Figure 3-6: Relationship between mass and velocity

WHAT IS GENERAL RELATIVITY?

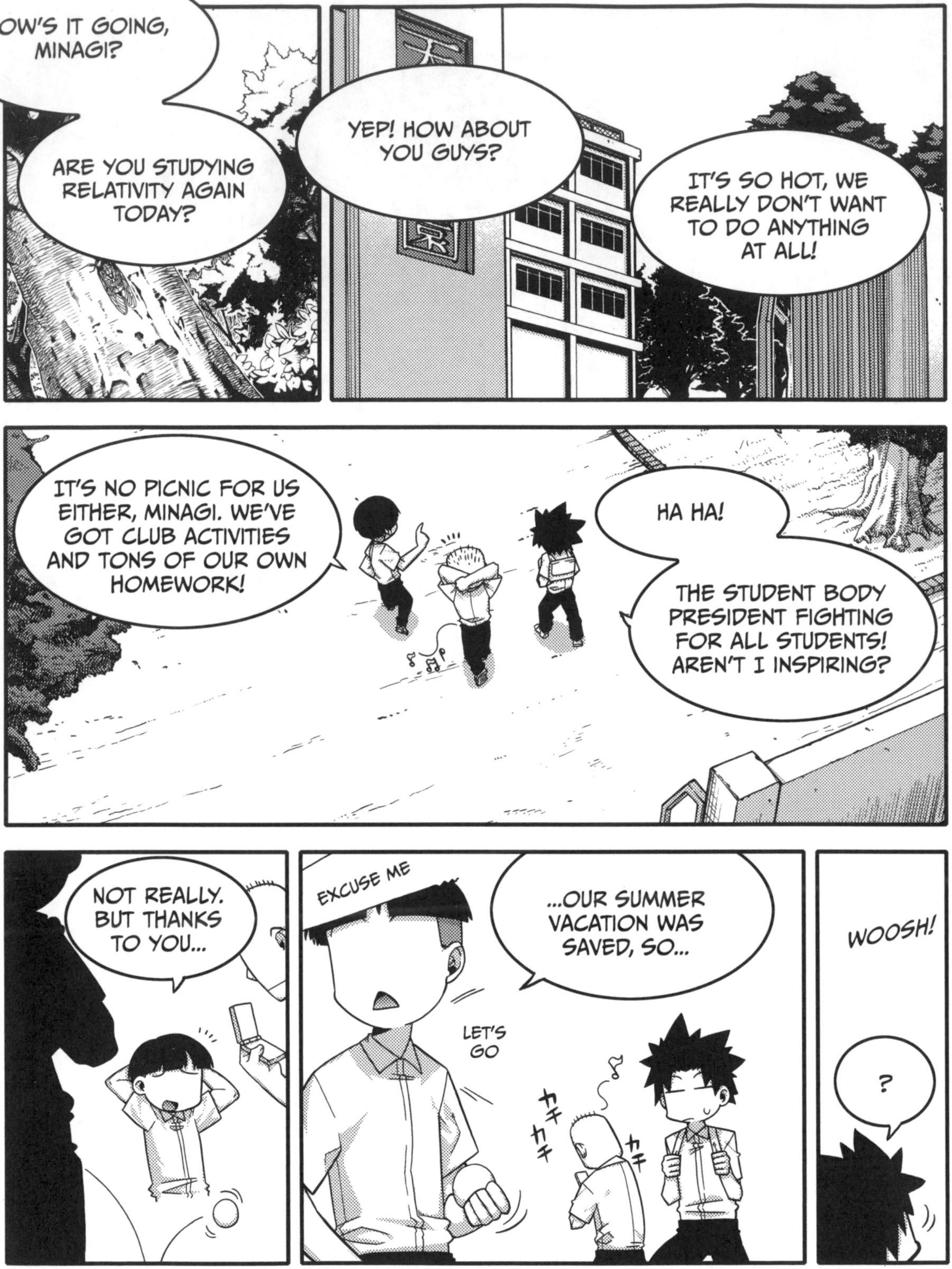
HOW'S IT GOING, MINAGI?
ARE YOU STUDYING RELATIVITY AGAIN TODAY?
YEP! HOW ABOUT YOU GUYS?
IT'S SO HOT, WE REALLY DON'T WANT TO DO ANYTHING AT ALL!
IT'S NO PICNIC FOR US EITHER, MINAGI. WE'VE GOT CLUB ACTIVITIES AND TONS OF OUR OWN HOMEWORK!
HA HA!
THE STUDENT BODY PRESIDENT FIGHTING FOR ALL STUDENTS! AREN'T I INSPIRING?
NOT REALLY. BUT THANKS TO YOU...
EXCUSE ME
...OUR SUMMER VACATION WAS SAVED, SO...
LET'S GO
カチ
カチ
カチ
WOOSH!
?

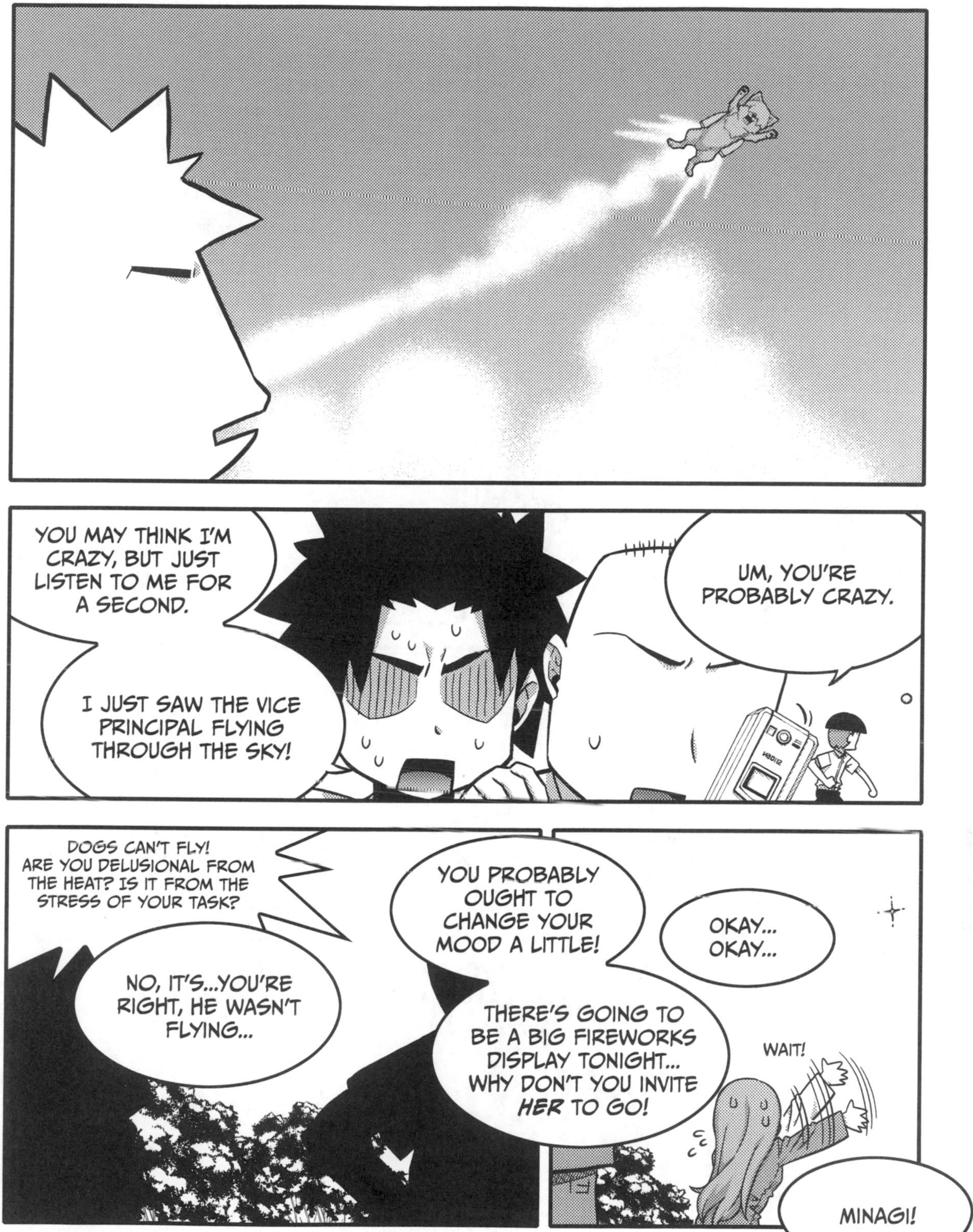
YOU MAY THINK I'M CRAZY, BUT JUST LISTEN TO ME FOR A SECOND.
I JUST SAW THE VICE PRINCIPAL FLYING THROUGH THE SKY!
UM, YOU'RE PROBABLY CRAZY.
DOGS CAN'T FLY! ARE YOU DELUSIONAL FROM THE HEAT? IS IT FROM THE STRESS OF YOUR TASK?
NO, IT'S...YOU'RE RIGHT, HE WASN'T FLYING...
YOU PROBABLY OUGHT TO CHANGE YOUR MOOD A LITTLE!
OKAY... OKAY...
THERE'S GOING TO BE A BIG FIREWORKS DISPLAY TONIGHT... WHY DON'T YOU INVITE *HER* TO GO!
WAIT!
MINAGI!

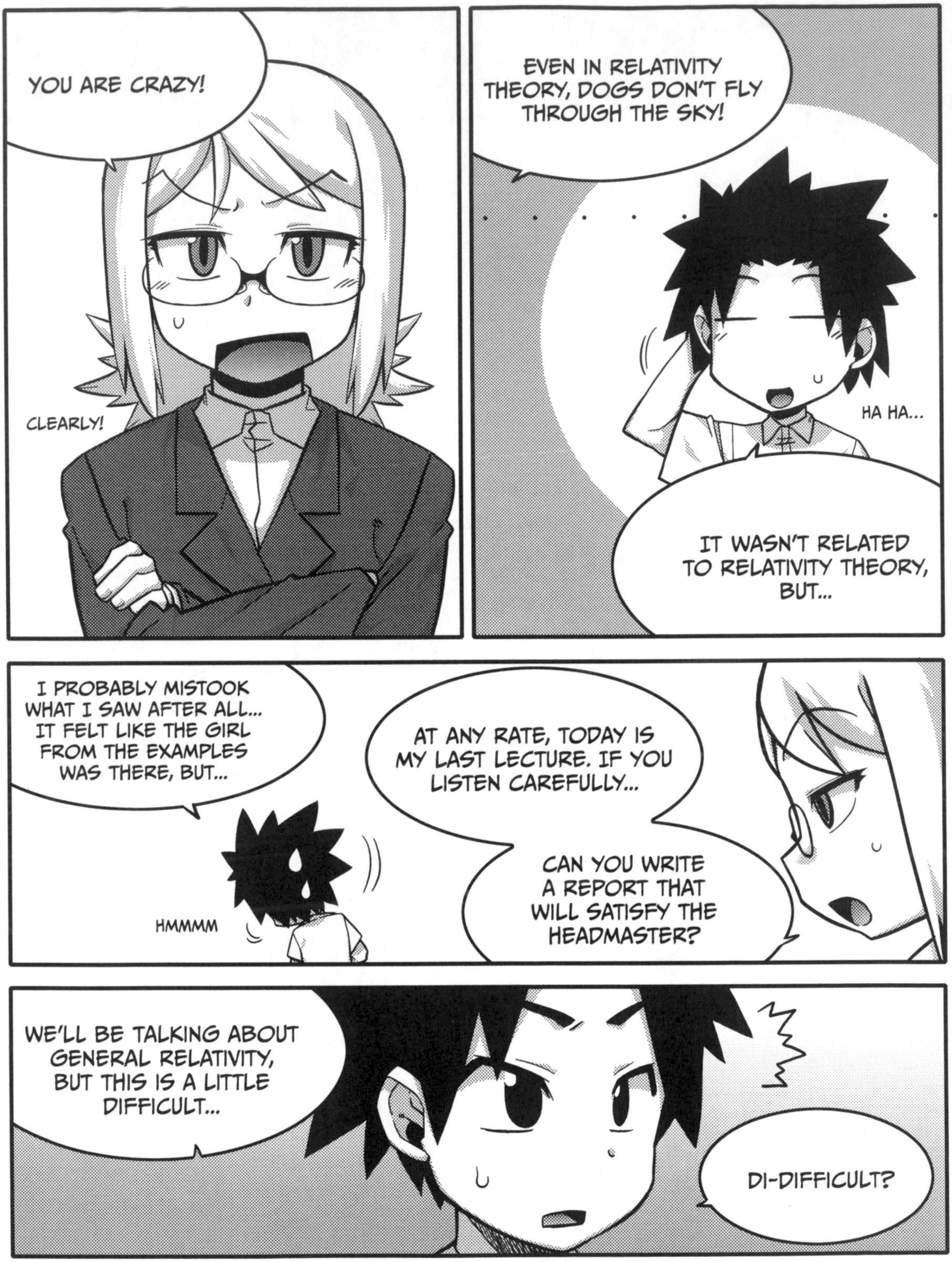

YOU ARE CRAZY!
CLEARLY!
EVEN IN RELATIVITY THEORY, DOGS DON'T FLY THROUGH THE SKY!
HA HA...
IT WASN'T RELATED TO RELATIVITY THEORY, BUT...
I PROBABLY MISTOOK WHAT I SAW AFTER ALL... IT FELT LIKE THE GIRL FROM THE EXAMPLES WAS THERE, BUT...
HMMMM
AT ANY RATE, TODAY IS MY LAST LECTURE. IF YOU LISTEN CAREFULLY...
CAN YOU WRITE A REPORT THAT WILL SATISFY THE HEADMASTER?
WE'LL BE TALKING ABOUT GENERAL RELATIVITY, BUT THIS IS A LITTLE DIFFICULT...
DI-DIFFICULT?

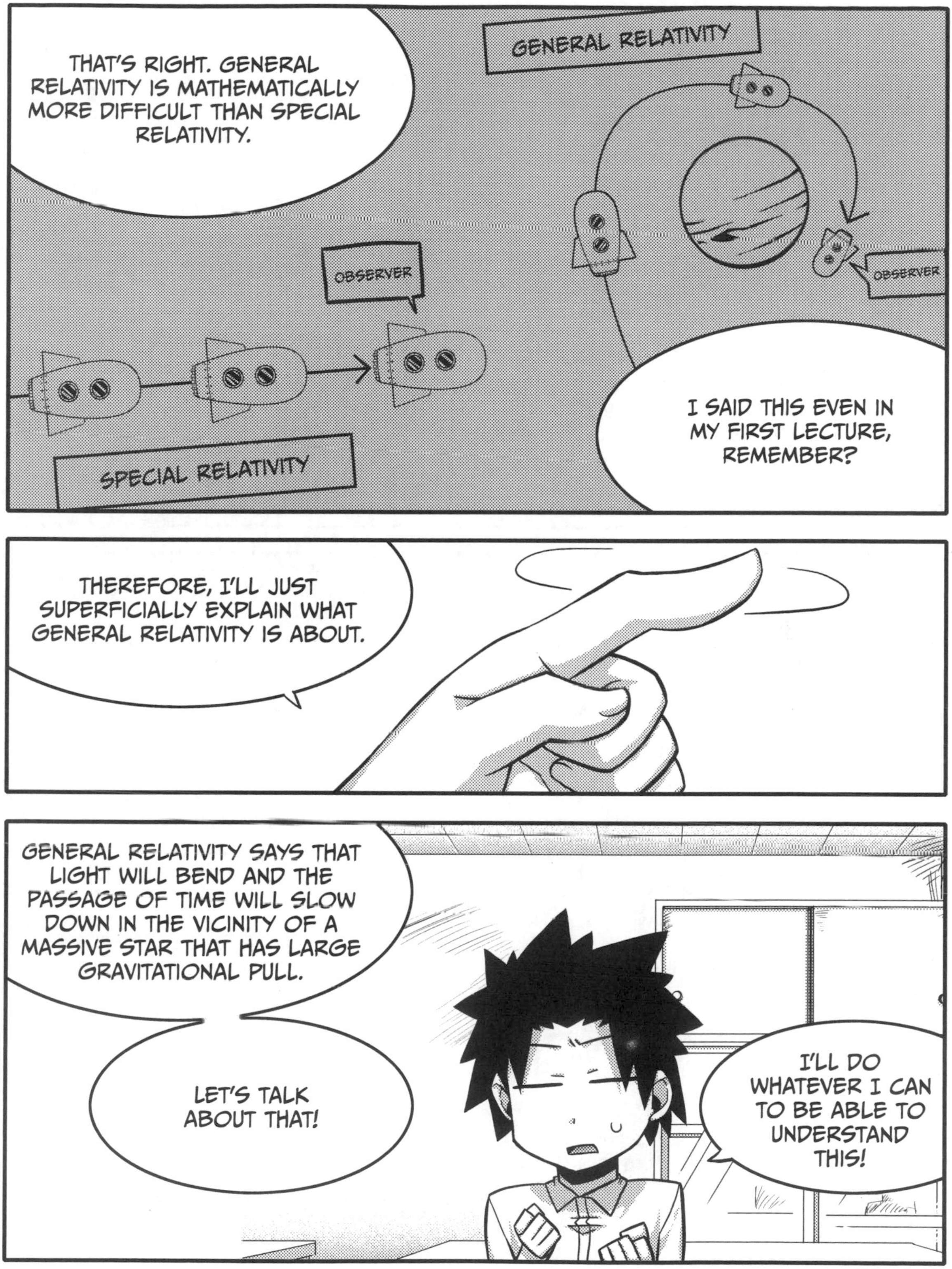
THAT'S RIGHT. GENERAL RELATIVITY IS MATHEMATICALLY MORE DIFFICULT THAN SPECIAL RELATIVITY.
GENERAL RELATIVITY
OBSERVER
OBSERVER
SPECIAL RELATIVITY
I SAID THIS EVEN IN MY FIRST LECTURE, REMEMBER?
THEREFORE, I'LL JUST SUPERFICIALLY EXPLAIN WHAT GENERAL RELATIVITY IS ABOUT.
GENERAL RELATIVITY SAYS THAT LIGHT WILL BEND AND THE PASSAGE OF TIME WILL SLOW DOWN IN THE VICINITY OF A MASSIVE STAR THAT HAS LARGE GRAVITATIONAL PULL.
LET'S TALK ABOUT THAT!
I'LL DO WHATEVER I CAN TO BE ABLE TO UNDERSTAND THIS!

1. EQUIVALENCE PRINCIPLE

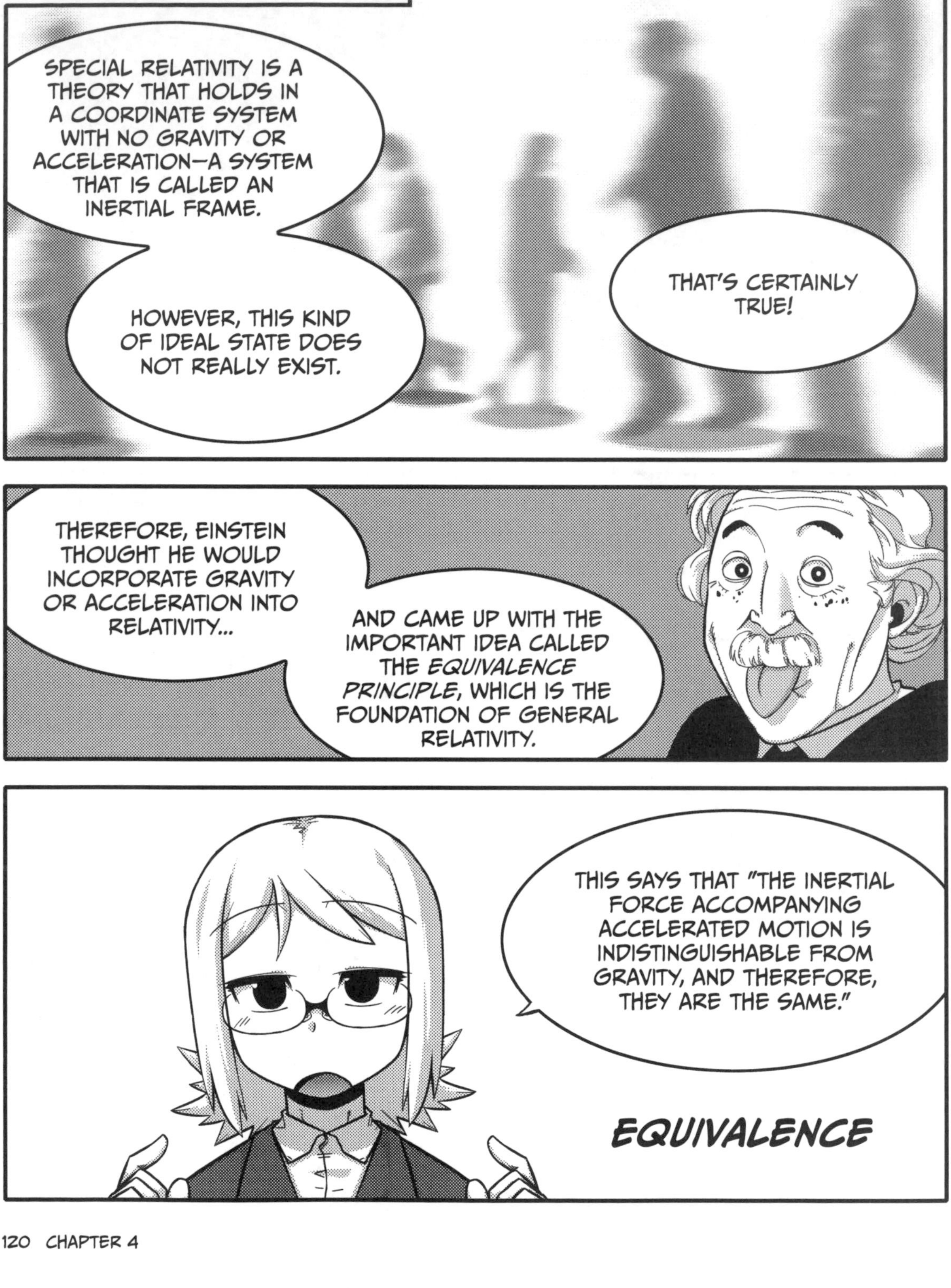
SPECIAL RELATIVITY IS A THEORY THAT HOLDS IN A COORDINATE SYSTEM WITH NO GRAVITY OR ACCELERATION—A SYSTEM THAT IS CALLED AN INERTIAL FRAME.
HOWEVER, THIS KIND OF IDEAL STATE DOES NOT REALLY EXIST.
THAT'S CERTAINLY TRUE!
THEREFORE, EINSTEIN THOUGHT HE WOULD INCORPORATE GRAVITY OR ACCELERATION INTO RELATIVITY...
AND CAME UP WITH THE IMPORTANT IDEA CALLED THE *EQUIVALENCE PRINCIPLE*, WHICH IS THE FOUNDATION OF GENERAL RELATIVITY.
THIS SAYS THAT "THE INERTIAL FORCE ACCOMPANYING ACCELERATED MOTION IS INDISTINGUISHABLE FROM GRAVITY, AND THEREFORE, THEY ARE THE SAME."
EQUIVALENCE

I DON'T REALLY UNDERSTAND INERTIAL FORCE...

I'LL EXPLAIN WHAT IT IS.

FIRST, ASSUME THAT YOU, MINAGI, ARE RIDING IN A ROCKET THAT IS FLOATING IN OUTER SPACE.

THERE'S NOTHING HERE...

THIS ROCKET HAS NO WINDOWS, AND YOU DON'T KNOW WHAT'S GOING ON OUTSIDE OF IT.

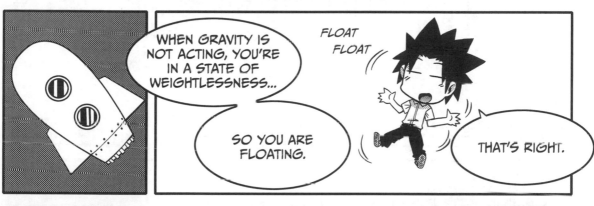

WHEN GRAVITY IS NOT ACTING, YOU'RE IN A STATE OF WEIGHTLESSNESS...

SO YOU ARE FLOATING.

FLOAT FLOAT

THAT'S RIGHT.

HOWEVER, IF THE ROCKET ACCELERATES...

YOUR BODY WILL BE PUSHED IN THE OPPOSITE DIRECTION TO THE ROCKET'S DIRECTION OF MOTION.

ERRRGH!

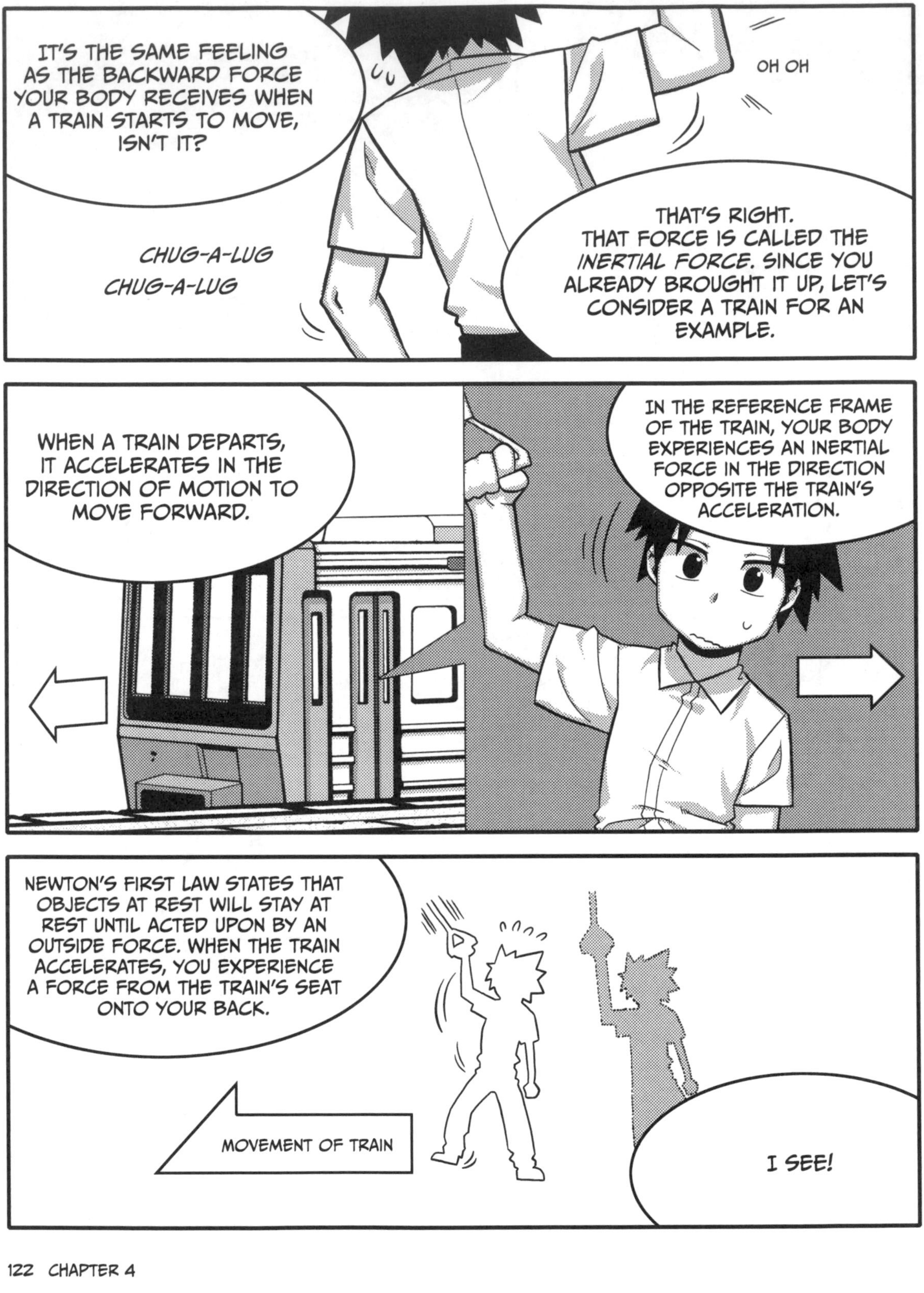
IT'S THE SAME FEELING AS THE BACKWARD FORCE YOUR BODY RECEIVES WHEN A TRAIN STARTS TO MOVE, ISN'T IT?
OH OH
CHUG-A-LUG
CHUG-A-LUG
THAT'S RIGHT. THAT FORCE IS CALLED THE *INERTIAL FORCE*. SINCE YOU ALREADY BROUGHT IT UP, LET'S CONSIDER A TRAIN FOR AN EXAMPLE.
WHEN A TRAIN DEPARTS, IT ACCELERATES IN THE DIRECTION OF MOTION TO MOVE FORWARD.
IN THE REFERENCE FRAME OF THE TRAIN, YOUR BODY EXPERIENCES AN INERTIAL FORCE IN THE DIRECTION OPPOSITE THE TRAIN'S ACCELERATION.
NEWTON'S FIRST LAW STATES THAT OBJECTS AT REST WILL STAY AT REST UNTIL ACTED UPON BY AN OUTSIDE FORCE. WHEN THE TRAIN ACCELERATES, YOU EXPERIENCE A FORCE FROM THE TRAIN'S SEAT ONTO YOUR BACK.
MOVEMENT OF TRAIN
I SEE!

IN OTHER WORDS, THIS HAPPENS BECAUSE YOUR BODY OBEYS THE LAW OF INERTIA.

THE FORCE YOU FEEL AT THAT TIME IS CALLED THE *INERTIAL FORCE*.

INERTIAL FORCE

THIS INERTIAL FORCE IS FELT ONLY BY PEOPLE WHO ARE INSIDE THE TRAIN.

OHHHHMIGOSH...

?

CREEEAK

THIS INERTIAL FORCE IS NOT FELT BY A PERSON ON THE PLATFORM OUTSIDE OF THE TRAIN.

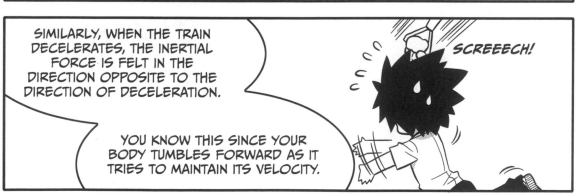

SIMILARLY, WHEN THE TRAIN DECELERATES, THE INERTIAL FORCE IS FELT IN THE DIRECTION OPPOSITE TO THE DIRECTION OF DECELERATION.

SCREEECH!

YOU KNOW THIS SINCE YOUR BODY TUMBLES FORWARD AS IT TRIES TO MAINTAIN ITS VELOCITY.

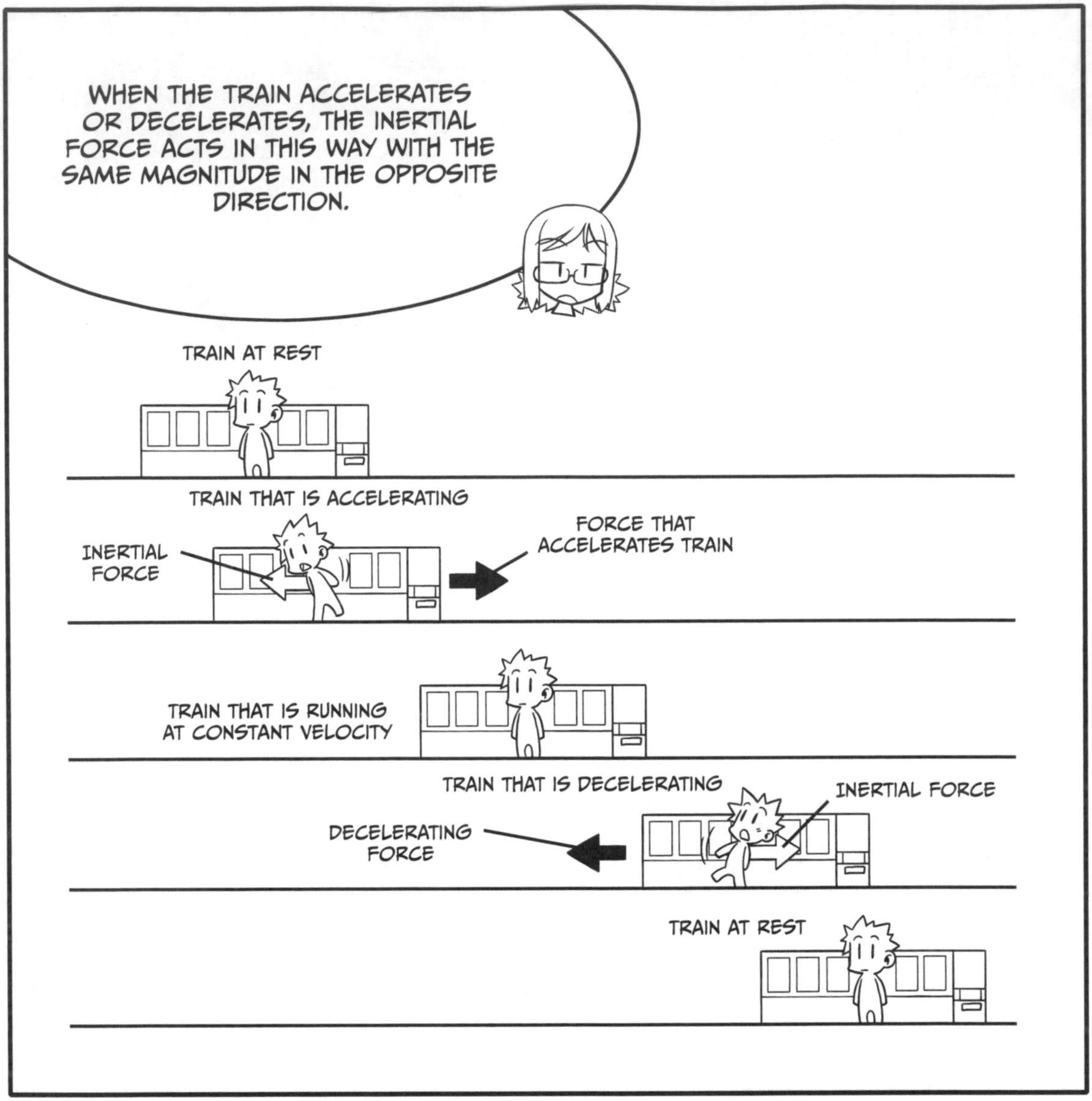
WHEN THE TRAIN ACCELERATES OR DECELERATES, THE INERTIAL FORCE ACTS IN THIS WAY WITH THE SAME MAGNITUDE IN THE OPPOSITE DIRECTION.
TRAIN AT REST
TRAIN THAT IS ACCELERATING
INERTIAL FORCE
FORCE THAT ACCELERATES TRAIN
TRAIN THAT IS RUNNING AT CONSTANT VELOCITY
TRAIN THAT IS DECELERATING
INERTIAL FORCE
DECELERATING FORCE
TRAIN AT REST

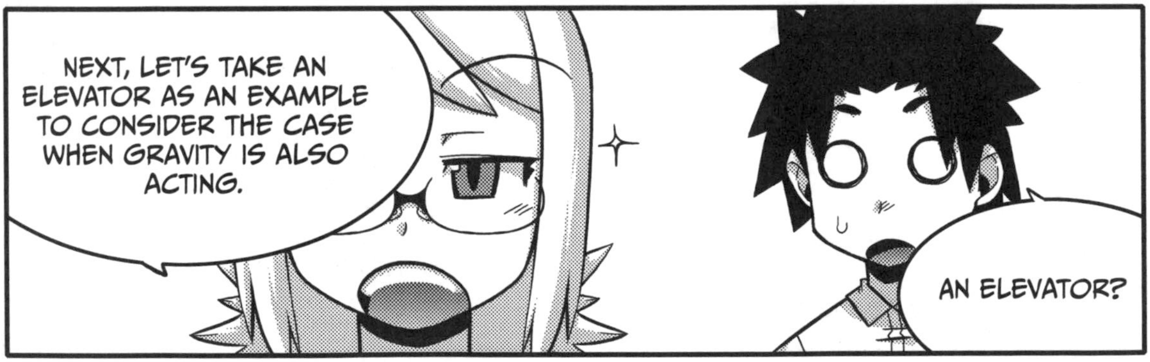
NEXT, LET'S TAKE AN ELEVATOR AS AN EXAMPLE TO CONSIDER THE CASE WHEN GRAVITY IS ALSO ACTING.
AN ELEVATOR?

WHEN THE ELEVATOR IS STOPPED AT THE BOTTOM, ONLY GRAVITY IS ACTING ON THE PEOPLE INSIDE.

GOING UP?

THEREFORE, THE PEOPLE INSIDE FEEL THEIR NORMAL BODY WEIGHT.

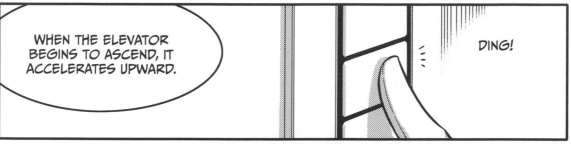

WHEN THE ELEVATOR BEGINS TO ASCEND, IT ACCELERATES UPWARD.

DING!

AT THIS TIME, THE PEOPLE INSIDE FEEL A FORCE IN THE DIRECTION OPPOSITE TO THE ELEVATOR'S DIRECTION OF ACCELERATION IN ADDITION TO GRAVITY.

THIS IS THE INERTIAL FORCE.

OF COURSE!

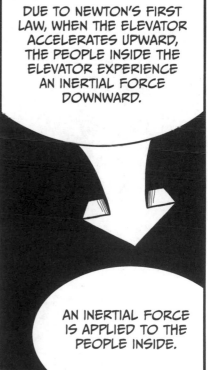

DUE TO NEWTON'S FIRST LAW, WHEN THE ELEVATOR ACCELERATES UPWARD, THE PEOPLE INSIDE THE ELEVATOR EXPERIENCE AN INERTIAL FORCE DOWNWARD.

AN INERTIAL FORCE IS APPLIED TO THE PEOPLE INSIDE.

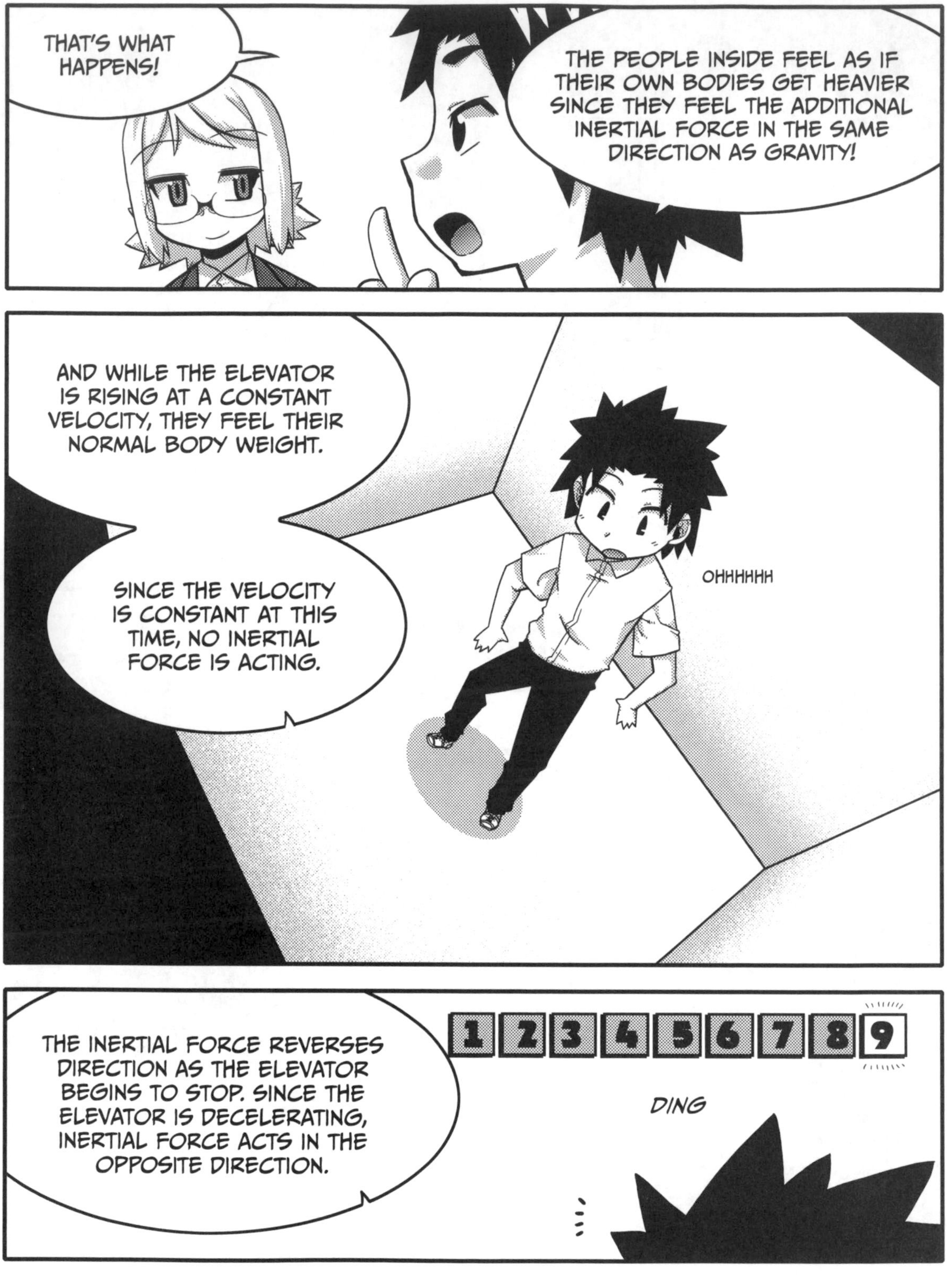
THAT'S WHAT HAPPENS!
THE PEOPLE INSIDE FEEL AS IF THEIR OWN BODIES GET HEAVIER SINCE THEY FEEL THE ADDITIONAL INERTIAL FORCE IN THE SAME DIRECTION AS GRAVITY!
AND WHILE THE ELEVATOR IS RISING AT A CONSTANT VELOCITY, THEY FEEL THEIR NORMAL BODY WEIGHT.
SINCE THE VELOCITY IS CONSTANT AT THIS TIME, NO INERTIAL FORCE IS ACTING.
OHHHHHH
THE INERTIAL FORCE REVERSES DIRECTION AS THE ELEVATOR BEGINS TO STOP. SINCE THE ELEVATOR IS DECELERATING, INERTIAL FORCE ACTS IN THE OPPOSITE DIRECTION.
1 2 3 4 5 6 7 8 9
DING

THE INERTIAL FORCE OFFSETS A PORTION OF GRAVITY'S FORCE.
THEREFORE, YOU FEEL AS IF YOU GET LIGHTER, DON'T YOU?
FLOAT
DO YOU UNDERSTAND THAT THIS ELEVATOR EXAMPLE DEMONSTRATES THE SAME EFFECT AS OUR TRAIN? WE'VE SIMPLY ADDED GRAVITY (THAT IS, ANOTHER FORCE) TO THE MIX.

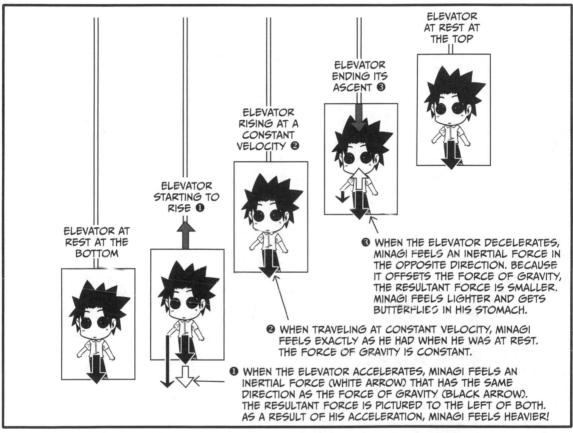

ELEVATOR AT REST AT THE BOTTOM
ELEVATOR STARTING TO RISE ❶
ELEVATOR RISING AT A CONSTANT VELOCITY ❷
ELEVATOR ENDING ITS ASCENT ❸
ELEVATOR AT REST AT THE TOP
❸ WHEN THE ELEVATOR DECELERATES, MINAGI FEELS AN INERTIAL FORCE IN THE OPPOSITE DIRECTION. BECAUSE IT OFFSETS THE FORCE OF GRAVITY, THE RESULTANT FORCE IS SMALLER. MINAGI FEELS LIGHTER AND GETS BUTTERFLIES IN HIS STOMACH.
❷ WHEN TRAVELING AT CONSTANT VELOCITY, MINAGI FEELS EXACTLY AS HE HAD WHEN HE WAS AT REST. THE FORCE OF GRAVITY IS CONSTANT.
❶ WHEN THE ELEVATOR ACCELERATES, MINAGI FEELS AN INERTIAL FORCE (WHITE ARROW) THAT HAS THE SAME DIRECTION AS THE FORCE OF GRAVITY (BLACK ARROW). THE RESULTANT FORCE IS PICTURED TO THE LEFT OF BOTH. AS A RESULT OF HIS ACCELERATION, MINAGI FEELS HEAVIER!

NOW LET'S CONSIDER *CENTRIFUGAL FORCE* AS ONE MORE EXAMPLE OF AN INERTIAL FORCE.
WE'VE SEEN THAT AN INERTIAL FORCE APPEARS WHEN ACCELERATION (CHANGE IN VELOCITY) OCCURS.

YOU'VE RIDDEN ON THE SWINGS AT AN AMUSEMENT PARK BEFORE, RIGHT?

WE'RE HERE!

YOU SIT IN A CHAIR AND FLY AROUND AND AROUND. YOU FEEL A FORCE EXERTED ON YOU THAT PUSHES YOU OUTWARD. THIS IS A *CENTRIFUGAL* FORCE, A SPECIFIC KIND OF INERTIAL FORCE.

IS THAT SO?

THE PERSON ON THE SWING HAS A VELOCITY THAT IS CONSTANTLY CHANGING DIRECTION, SUCH THAT THE PERSON TRAVELS IN A CIRCLE.

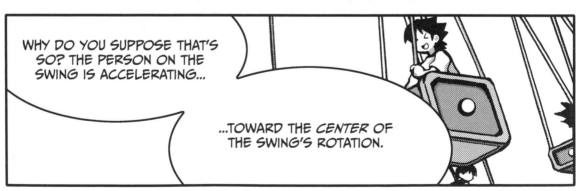

WHY DO YOU SUPPOSE THAT'S SO? THE PERSON ON THE SWING IS ACCELERATING...

...TOWARD THE *CENTER* OF THE SWING'S ROTATION.

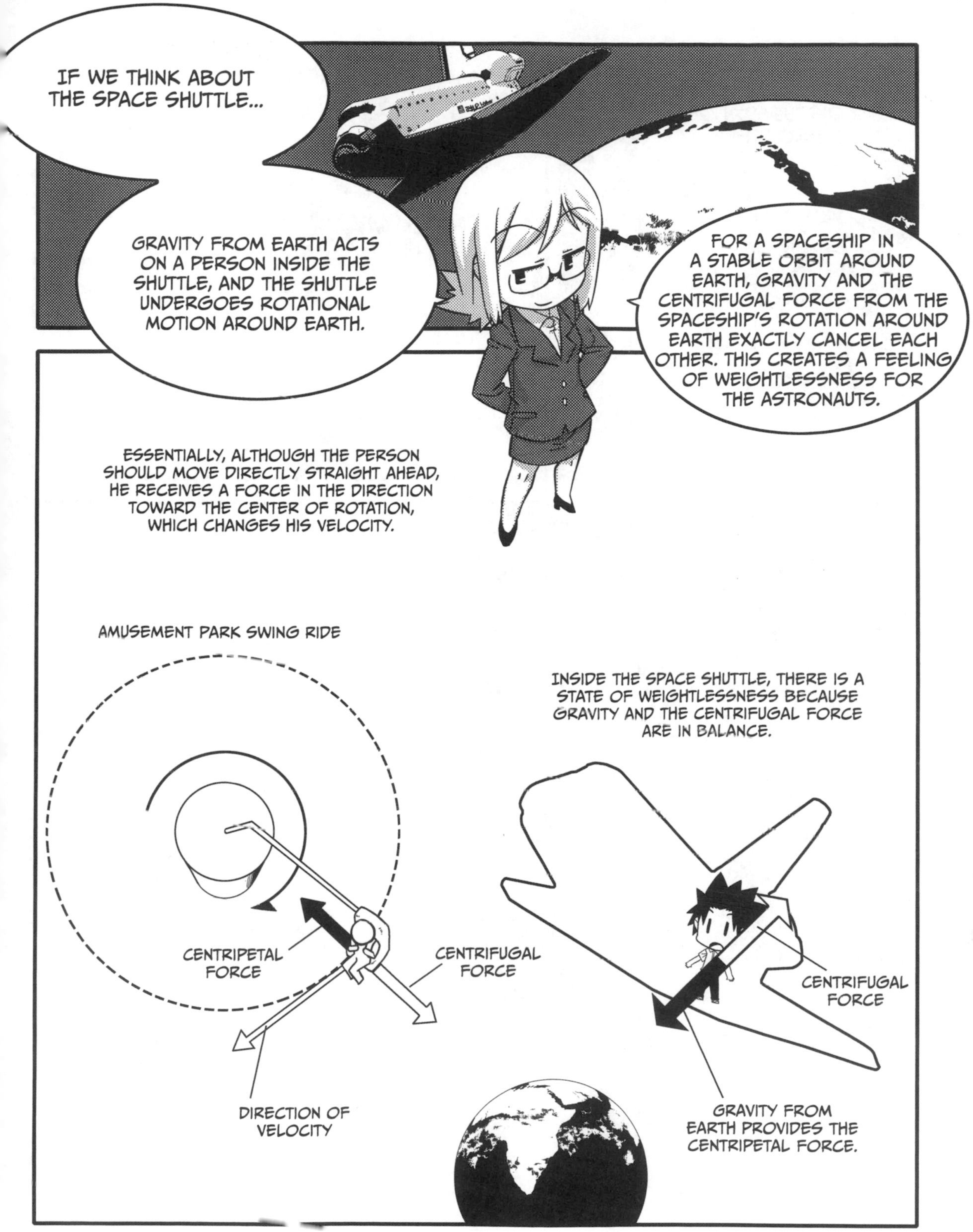
IF WE THINK ABOUT THE SPACE SHUTTLE...
GRAVITY FROM EARTH ACTS ON A PERSON INSIDE THE SHUTTLE, AND THE SHUTTLE UNDERGOES ROTATIONAL MOTION AROUND EARTH.
FOR A SPACESHIP IN A STABLE ORBIT AROUND EARTH, GRAVITY AND THE CENTRIFUGAL FORCE FROM THE SPACESHIP'S ROTATION AROUND EARTH EXACTLY CANCEL EACH OTHER. THIS CREATES A FEELING OF WEIGHTLESSNESS FOR THE ASTRONAUTS.
ESSENTIALLY, ALTHOUGH THE PERSON SHOULD MOVE DIRECTLY STRAIGHT AHEAD, HE RECEIVES A FORCE IN THE DIRECTION TOWARD THE CENTER OF ROTATION, WHICH CHANGES HIS VELOCITY.
AMUSEMENT PARK SWING RIDE
CENTRIPETAL FORCE
CENTRIFUGAL FORCE
DIRECTION OF VELOCITY
INSIDE THE SPACE SHUTTLE, THERE IS A STATE OF WEIGHTLESSNESS BECAUSE GRAVITY AND THE CENTRIFUGAL FORCE ARE IN BALANCE.
CENTRIFUGAL FORCE
GRAVITY FROM EARTH PROVIDES THE CENTRIPETAL FORCE.

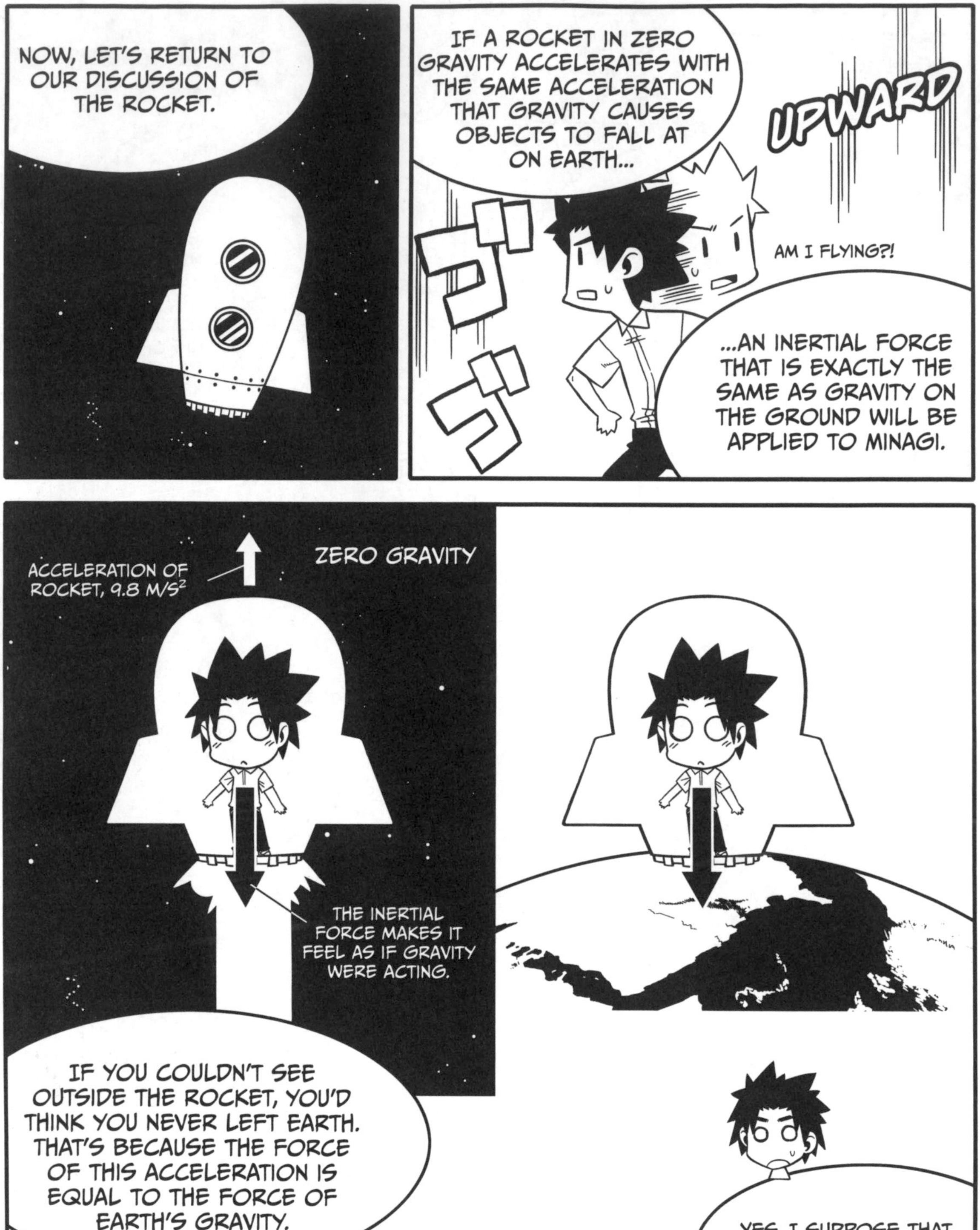
NOW, LET'S RETURN TO OUR DISCUSSION OF THE ROCKET.
IF A ROCKET IN ZERO GRAVITY ACCELERATES WITH THE SAME ACCELERATION THAT GRAVITY CAUSES OBJECTS TO FALL AT ON EARTH...
UPWARD
AM I FLYING?!
...AN INERTIAL FORCE THAT IS EXACTLY THE SAME AS GRAVITY ON THE GROUND WILL BE APPLIED TO MINAGI.
ACCELERATION OF ROCKET, 9.8 M/S²
ZERO GRAVITY
THE INERTIAL FORCE MAKES IT FEEL AS IF GRAVITY WERE ACTING.
IF YOU COULDN'T SEE OUTSIDE THE ROCKET, YOU'D THINK YOU NEVER LEFT EARTH. THAT'S BECAUSE THE FORCE OF THIS ACCELERATION IS EQUAL TO THE FORCE OF EARTH'S GRAVITY.
YES, I SUPPOSE THAT MAKES SENSE.

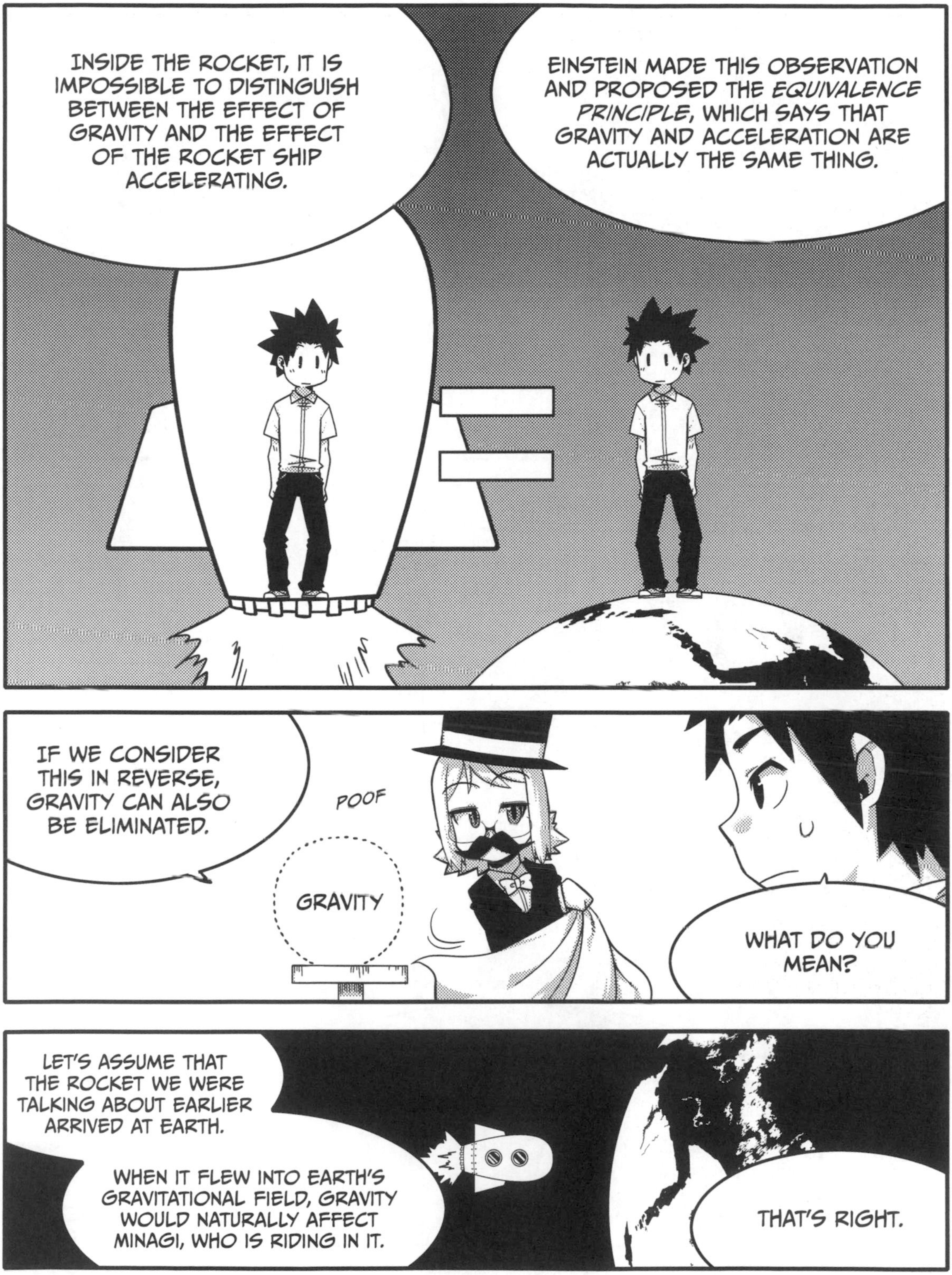
INSIDE THE ROCKET, IT IS IMPOSSIBLE TO DISTINGUISH BETWEEN THE EFFECT OF GRAVITY AND THE EFFECT OF THE ROCKET SHIP ACCELERATING.
EINSTEIN MADE THIS OBSERVATION AND PROPOSED THE *EQUIVALENCE PRINCIPLE*, WHICH SAYS THAT GRAVITY AND ACCELERATION ARE ACTUALLY THE SAME THING.
IF WE CONSIDER THIS IN REVERSE, GRAVITY CAN ALSO BE ELIMINATED.
POOF
GRAVITY
WHAT DO YOU MEAN?
LET'S ASSUME THAT THE ROCKET WE WERE TALKING ABOUT EARLIER ARRIVED AT EARTH.
WHEN IT FLEW INTO EARTH'S GRAVITATIONAL FIELD, GRAVITY WOULD NATURALLY AFFECT MINAGI, WHO IS RIDING IN IT.
THAT'S RIGHT.

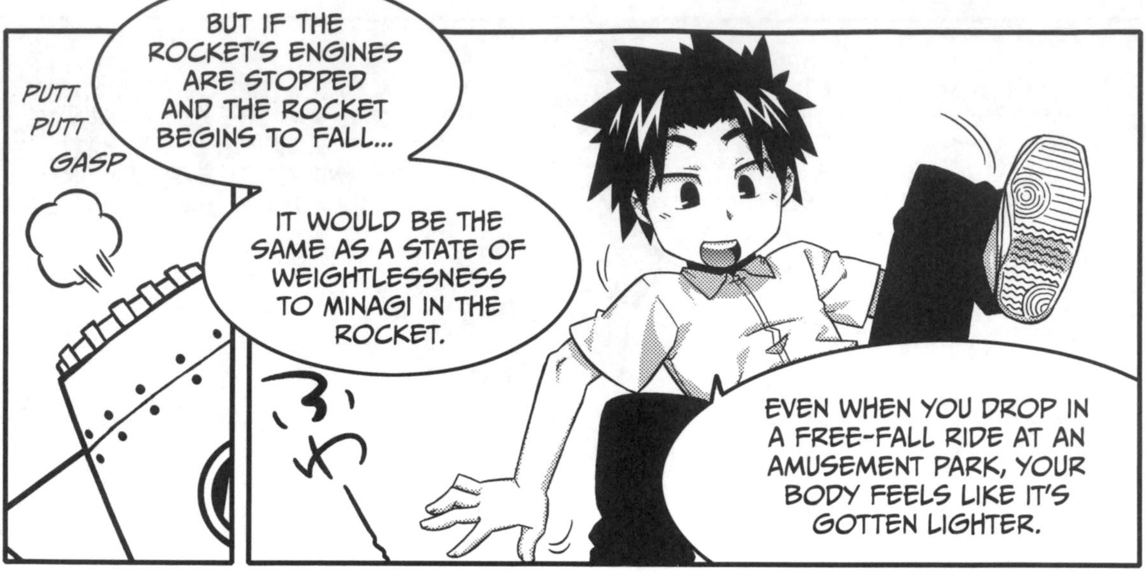
PUTT
PUTT
GASP
BUT IF THE ROCKET'S ENGINES ARE STOPPED AND THE ROCKET BEGINS TO FALL...
IT WOULD BE THE SAME AS A STATE OF WEIGHTLESSNESS TO MINAGI IN THE ROCKET.
EVEN WHEN YOU DROP IN A FREE-FALL RIDE AT AN AMUSEMENT PARK, YOUR BODY FEELS LIKE IT'S GOTTEN LIGHTER.

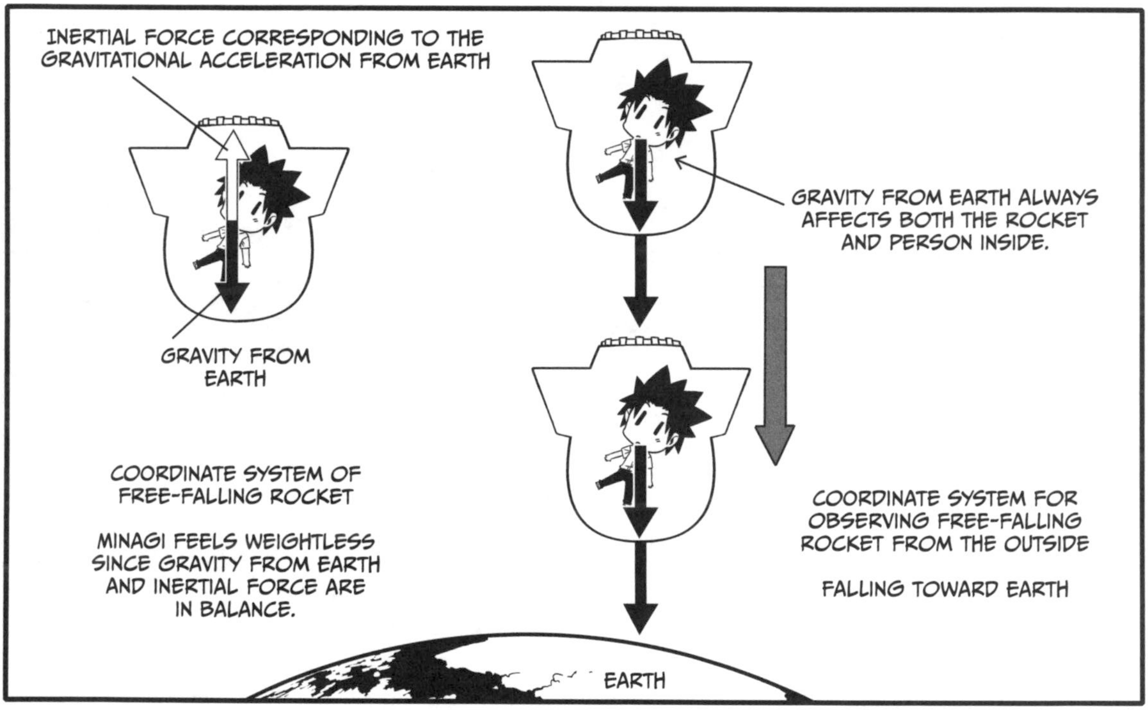
INERTIAL FORCE CORRESPONDING TO THE GRAVITATIONAL ACCELERATION FROM EARTH
GRAVITY FROM EARTH
GRAVITY FROM EARTH ALWAYS AFFECTS BOTH THE ROCKET AND PERSON INSIDE.
COORDINATE SYSTEM OF FREE-FALLING ROCKET
MINAGI FEELS WEIGHTLESS SINCE GRAVITY FROM EARTH AND INERTIAL FORCE ARE IN BALANCE.
COORDINATE SYSTEM FOR OBSERVING FREE-FALLING ROCKET FROM THE OUTSIDE
FALLING TOWARD EARTH
EARTH

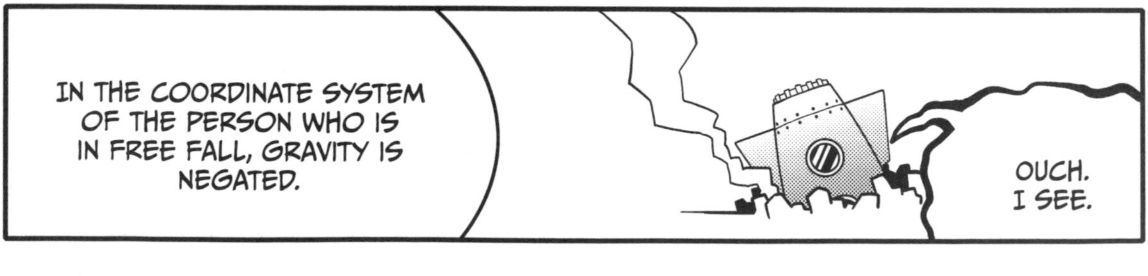
IN THE COORDINATE SYSTEM OF THE PERSON WHO IS IN FREE FALL, GRAVITY IS NEGATED.
OUCH. I SEE.

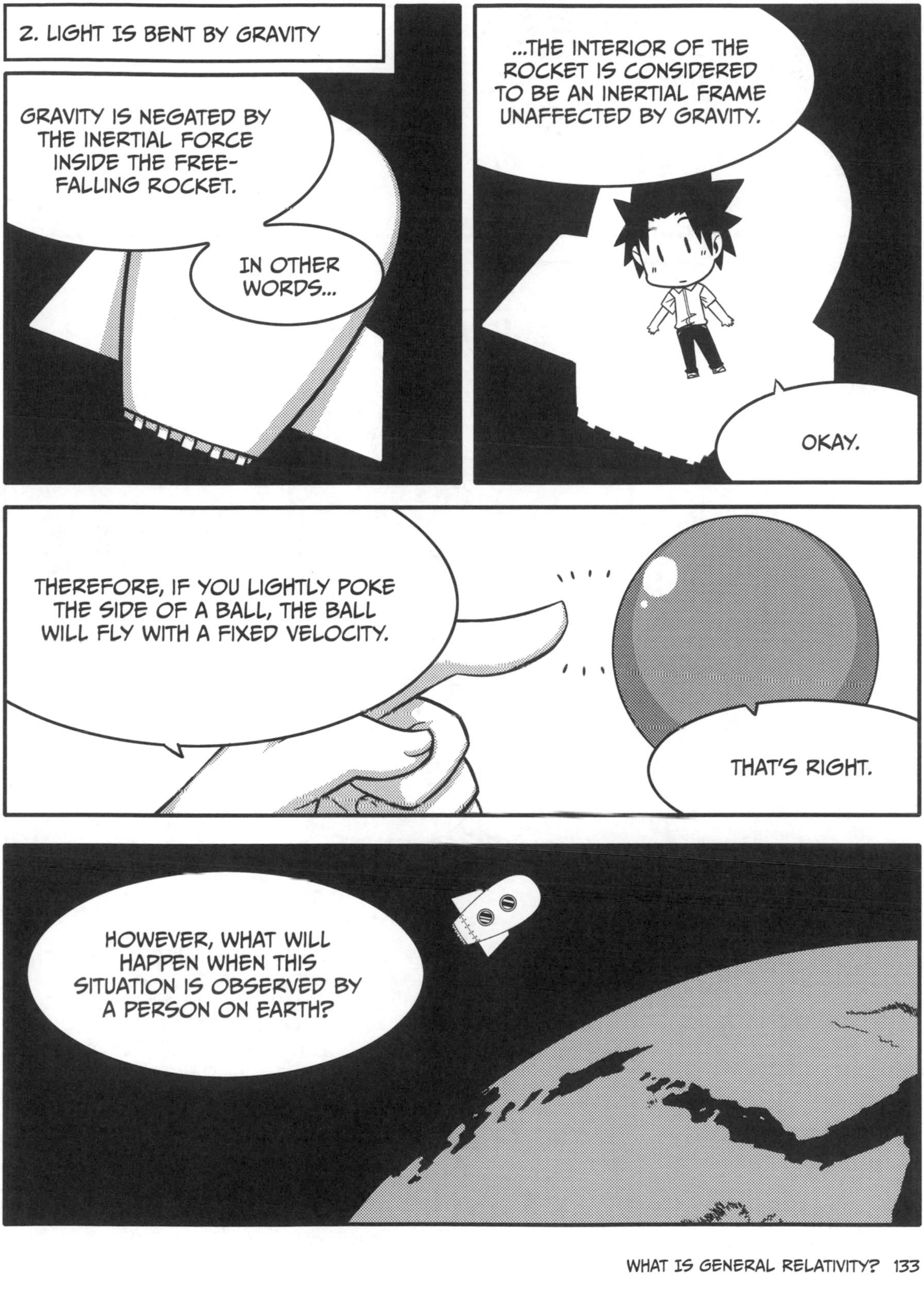
2. LIGHT IS BENT BY GRAVITY
GRAVITY IS NEGATED BY THE INERTIAL FORCE INSIDE THE FREE-FALLING ROCKET.
IN OTHER WORDS...
...THE INTERIOR OF THE ROCKET IS CONSIDERED TO BE AN INERTIAL FRAME UNAFFECTED BY GRAVITY.
OKAY.
THEREFORE, IF YOU LIGHTLY POKE THE SIDE OF A BALL, THE BALL WILL FLY WITH A FIXED VELOCITY.
THAT'S RIGHT.
HOWEVER, WHAT WILL HAPPEN WHEN THIS SITUATION IS OBSERVED BY A PERSON ON EARTH?

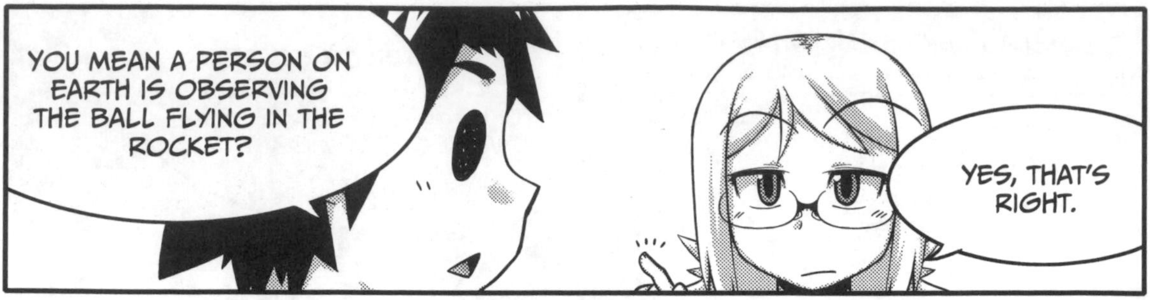
YOU MEAN A PERSON ON EARTH IS OBSERVING THE BALL FLYING IN THE ROCKET?
YES, THAT'S RIGHT.

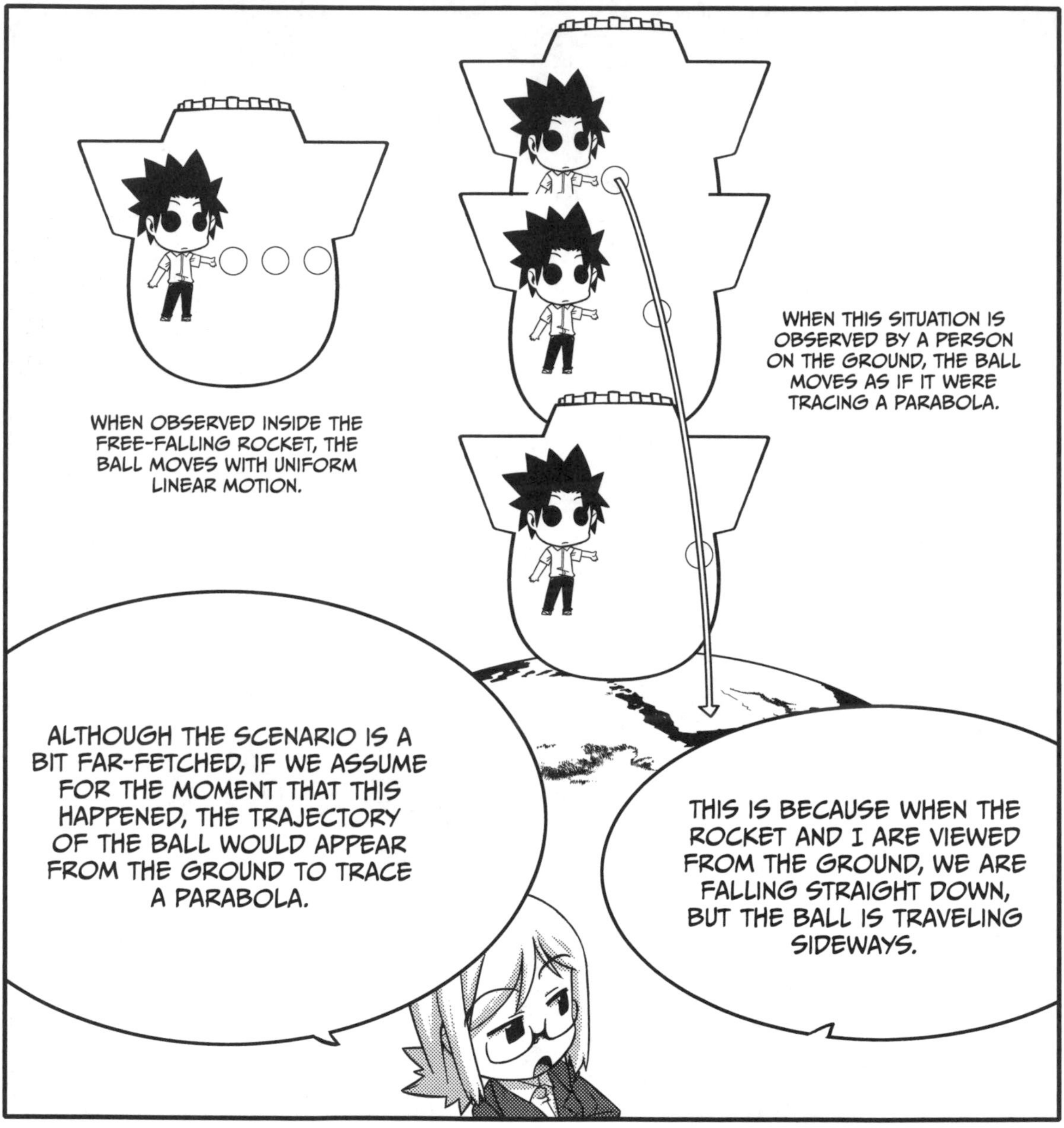
WHEN OBSERVED INSIDE THE FREE-FALLING ROCKET, THE BALL MOVES WITH UNIFORM LINEAR MOTION.
WHEN THIS SITUATION IS OBSERVED BY A PERSON ON THE GROUND, THE BALL MOVES AS IF IT WERE TRACING A PARABOLA.
ALTHOUGH THE SCENARIO IS A BIT FAR-FETCHED, IF WE ASSUME FOR THE MOMENT THAT THIS HAPPENED, THE TRAJECTORY OF THE BALL WOULD APPEAR FROM THE GROUND TO TRACE A PARABOLA.
THIS IS BECAUSE WHEN THE ROCKET AND I ARE VIEWED FROM THE GROUND, WE ARE FALLING STRAIGHT DOWN, BUT THE BALL IS TRAVELING SIDEWAYS.

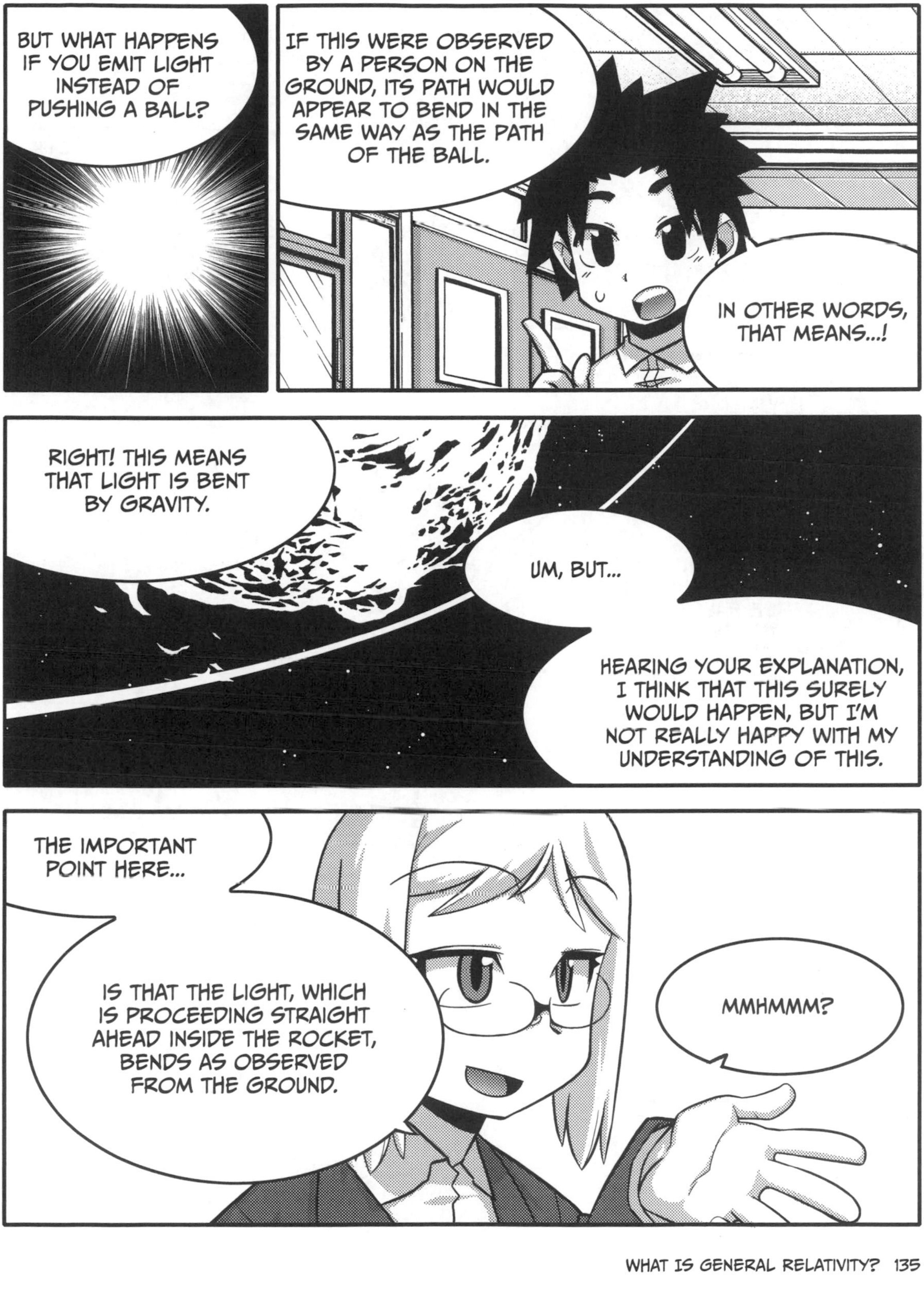
BUT WHAT HAPPENS IF YOU EMIT LIGHT INSTEAD OF PUSHING A BALL?
IF THIS WERE OBSERVED BY A PERSON ON THE GROUND, ITS PATH WOULD APPEAR TO BEND IN THE SAME WAY AS THE PATH OF THE BALL.
IN OTHER WORDS, THAT MEANS...!
RIGHT! THIS MEANS THAT LIGHT IS BENT BY GRAVITY.
UM, BUT...
HEARING YOUR EXPLANATION, I THINK THAT THIS SURELY WOULD HAPPEN, BUT I'M NOT REALLY HAPPY WITH MY UNDERSTANDING OF THIS.
THE IMPORTANT POINT HERE...
IS THAT THE LIGHT, WHICH IS PROCEEDING STRAIGHT AHEAD INSIDE THE ROCKET, BENDS AS OBSERVED FROM THE GROUND.
MMHMMM?

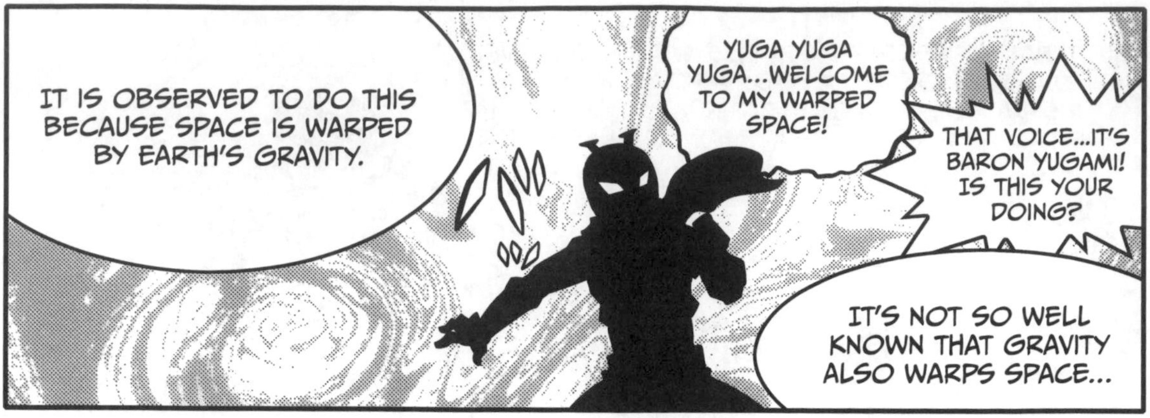
IT IS OBSERVED TO DO THIS BECAUSE SPACE IS WARPED BY EARTH'S GRAVITY.
YUGA YUGA YUGA...WELCOME TO MY WARPED SPACE!
THAT VOICE...IT'S BARON YUGAMI! IS THIS YOUR DOING?
IT'S NOT SO WELL KNOWN THAT GRAVITY ALSO WARPS SPACE...

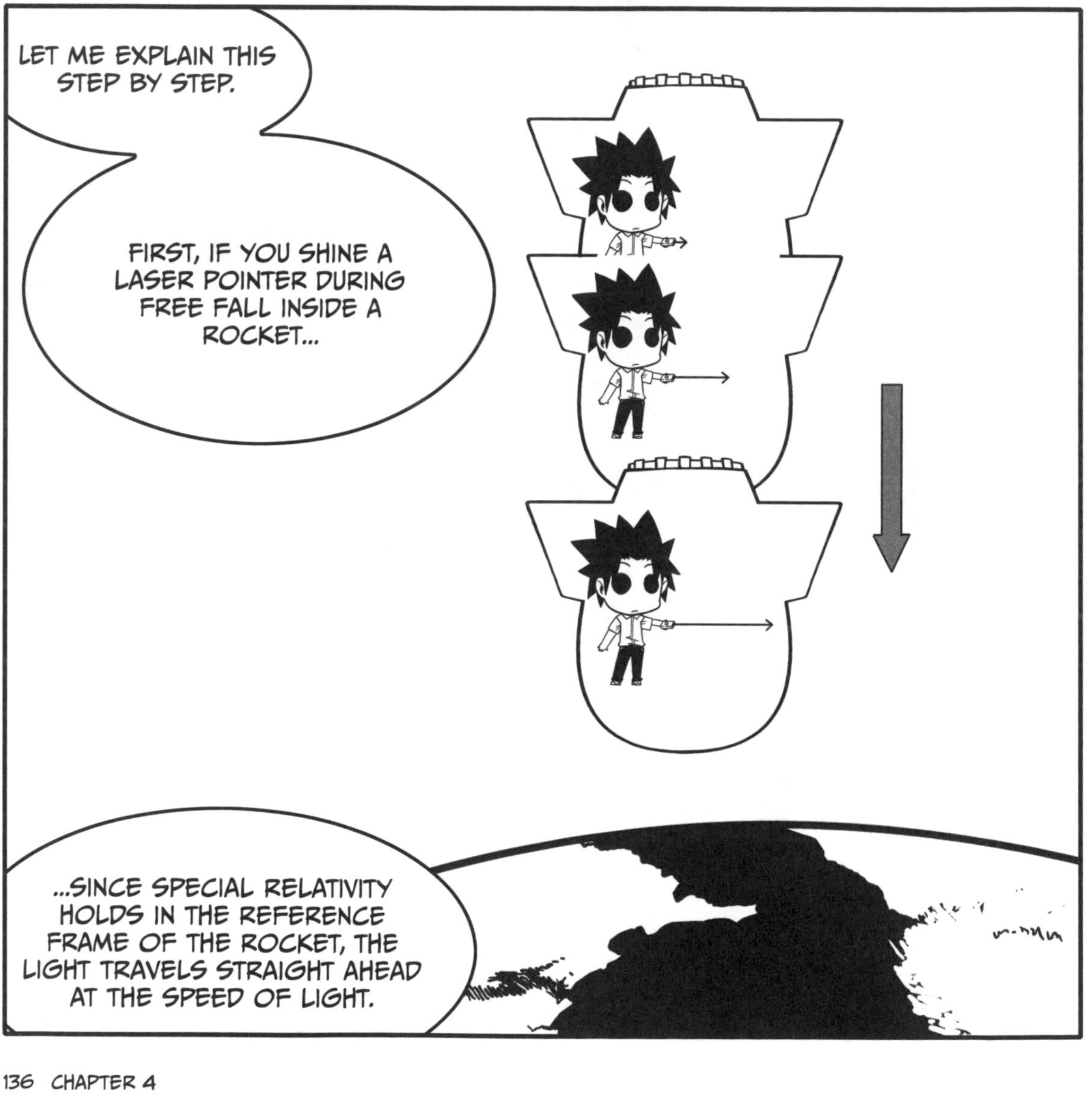
LET ME EXPLAIN THIS STEP BY STEP.
FIRST, IF YOU SHINE A LASER POINTER DURING FREE FALL INSIDE A ROCKET...
...SINCE SPECIAL RELATIVITY HOLDS IN THE REFERENCE FRAME OF THE ROCKET, THE LIGHT TRAVELS STRAIGHT AHEAD AT THE SPEED OF LIGHT.

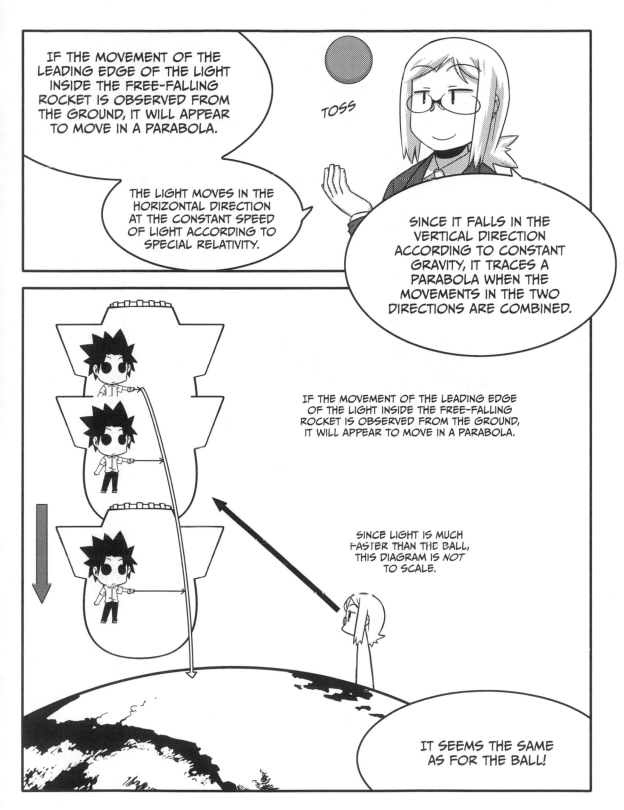

IF THE MOVEMENT OF THE LEADING EDGE OF THE LIGHT INSIDE THE FREE-FALLING ROCKET IS OBSERVED FROM THE GROUND, IT WILL APPEAR TO MOVE IN A PARABOLA.

TOSS

THE LIGHT MOVES IN THE HORIZONTAL DIRECTION AT THE CONSTANT SPEED OF LIGHT ACCORDING TO SPECIAL RELATIVITY.

SINCE IT FALLS IN THE VERTICAL DIRECTION ACCORDING TO CONSTANT GRAVITY, IT TRACES A PARABOLA WHEN THE MOVEMENTS IN THE TWO DIRECTIONS ARE COMBINED.

IF THE MOVEMENT OF THE LEADING EDGE OF THE LIGHT INSIDE THE FREE-FALLING ROCKET IS OBSERVED FROM THE GROUND, IT WILL APPEAR TO MOVE IN A PARABOLA.

SINCE LIGHT IS MUCH FASTER THAN THE BALL, THIS DIAGRAM IS *NOT* TO SCALE.

IT SEEMS THE SAME AS FOR THE BALL!

WHAT IS GENERAL RELATIVITY? 137

INSIDE THE FREE-FALLING ROCKET, THE LIGHT TRAVELS THE SHORTEST DISTANCE IN SPACE IN THE INERTIAL FRAME...OR, IN OTHER WORDS, IN A STRAIGHT LINE.

AND WHEN THE INTERIOR OF THE FREE-FALLING ROCKET IS OBSERVED FROM THE GROUND... THE LIGHT BENDS.

IT CERTAINLY APPEARS TO BE A ROUNDABOUT PATH THAT IS NOT TRAVELING ALONG THE SHORTEST DISTANCE.

DOESN'T THE WAY IN WHICH THE LIGHT TRAVELS APPEAR TO DIFFER WHEN IT IS OBSERVED FROM INSIDE THE ROCKET AND WHEN IT IS OBSERVED FROM THE GROUND?

YES, IT DOES, BUT...

GENERAL RELATIVITY ASSUMES THAT PHYSICAL PHENOMENA SUCH AS LIGHT TRAVELING ALONG THE SHORTEST DISTANCE IN SPACE-TIME...

SHOULD NOT DIFFER ACCORDING TO THE OBSERVATION POINT.

ISN'T THIS A CONTRADICTION?!

CALM DOWN.

SINCE SPACE-TIME IS FLAT IN AN INERTIAL FRAME, LIGHT IS OBSERVED TO PROCEED IN A STRAIGHT LINE...THIS IS WHAT HAPPENED, ISN'T IT?

UM...

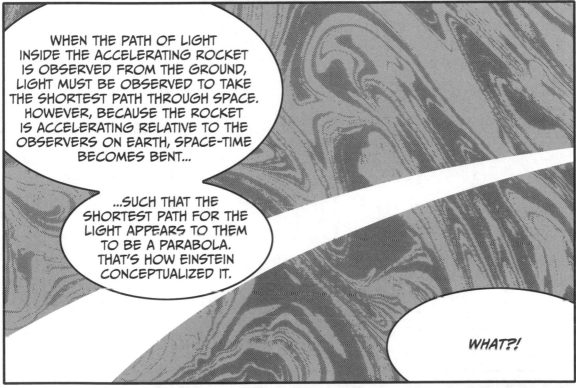

WHEN THE PATH OF LIGHT INSIDE THE ACCELERATING ROCKET IS OBSERVED FROM THE GROUND, LIGHT MUST BE OBSERVED TO TAKE THE SHORTEST PATH THROUGH SPACE. HOWEVER, BECAUSE THE ROCKET IS ACCELERATING RELATIVE TO THE OBSERVERS ON EARTH, SPACE-TIME BECOMES BENT...

...SUCH THAT THE SHORTEST PATH FOR THE LIGHT APPEARS TO THEM TO BE A PARABOLA. THAT'S HOW EINSTEIN CONCEPTUALIZED IT.

WHAT?!

WHAT AN OUTRAGEOUS CONCEPT.

IT SURE IS.

AND EINSTEIN WONDERED WHETHER THAT CURVATURE OF SPACE-TIME WAS GRAVITY ITSELF.

IT'S REALLY HARD TO IMAGINE CURVED SPACE-TIME.

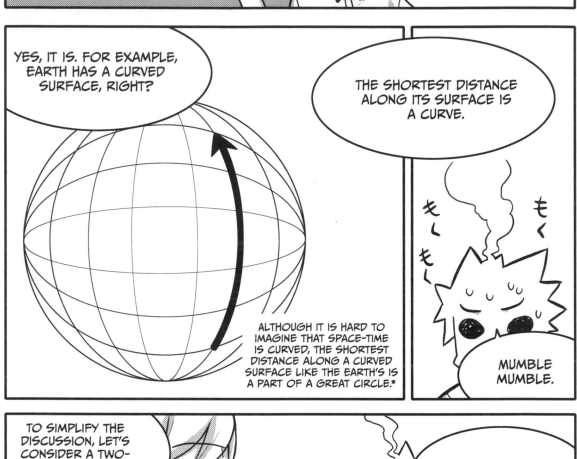

YES, IT IS. FOR EXAMPLE, EARTH HAS A CURVED SURFACE, RIGHT?

THE SHORTEST DISTANCE ALONG ITS SURFACE IS A CURVE.

ALTHOUGH IT IS HARD TO IMAGINE THAT SPACE-TIME IS CURVED, THE SHORTEST DISTANCE ALONG A CURVED SURFACE LIKE THE EARTH'S IS A PART OF A GREAT CIRCLE.*

MUMBLE MUMBLE.

TO SIMPLIFY THE DISCUSSION, LET'S CONSIDER A TWO-DIMENSIONAL SURFACE—THAT IS, A FLAT SURFACE.

THE ROUGH SKETCH COLLAPSES.

YES...LET'S!

* A GREAT CIRCLE IS ONE THAT EVENLY DIVIDES A SPHERE INTO TWO EQUAL HALVES. THE SHORTEST DISTANCE BETWEEN TWO POINTS ON A SPHERE IS *ALWAYS* A PORTION (OR A MINOR ARC) OF GREAT CIRCLE.

WHAT'S THIS?

IT'S A RUBBER SHEET STRETCHED FLAT LIKE A TRAMPOLINE. ASSUME THAT IT REPRESENTS *SPACE-TIME*.

NOW, LET'S PLACE A BOWLING BALL HERE.

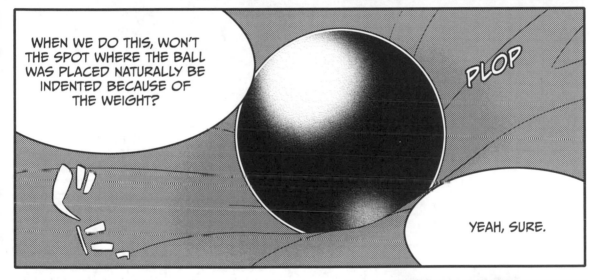

WHEN WE DO THIS, WON'T THE SPOT WHERE THE BALL WAS PLACED NATURALLY BE INDENTED BECAUSE OF THE WEIGHT?

PLOP

YEAH, SURE.

THIS INDENTATION IS A METAPHOR FOR THE WARPING OF SPACE.

WHAT?!

NOW LET'S ALSO PLACE ANOTHER BALL THERE.

WHEN WE DO THAT, THE SIZE OF THE INDENTATION WILL INCREASE, AND THE TWO BALLS WILL APPROACH EACH OTHER.

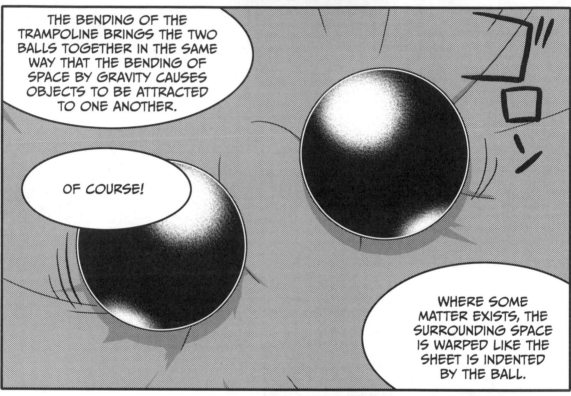

THE BENDING OF THE TRAMPOLINE BRINGS THE TWO BALLS TOGETHER IN THE SAME WAY THAT THE BENDING OF SPACE BY GRAVITY CAUSES OBJECTS TO BE ATTRACTED TO ONE ANOTHER.

OF COURSE!

WHERE SOME MATTER EXISTS, THE SURROUNDING SPACE IS WARPED LIKE THE SHEET IS INDENTED BY THE BALL.

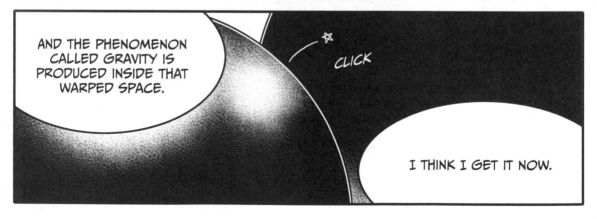

AND THE PHENOMENON CALLED GRAVITY IS PRODUCED INSIDE THAT WARPED SPACE.

CLICK

I THINK I GET IT NOW.

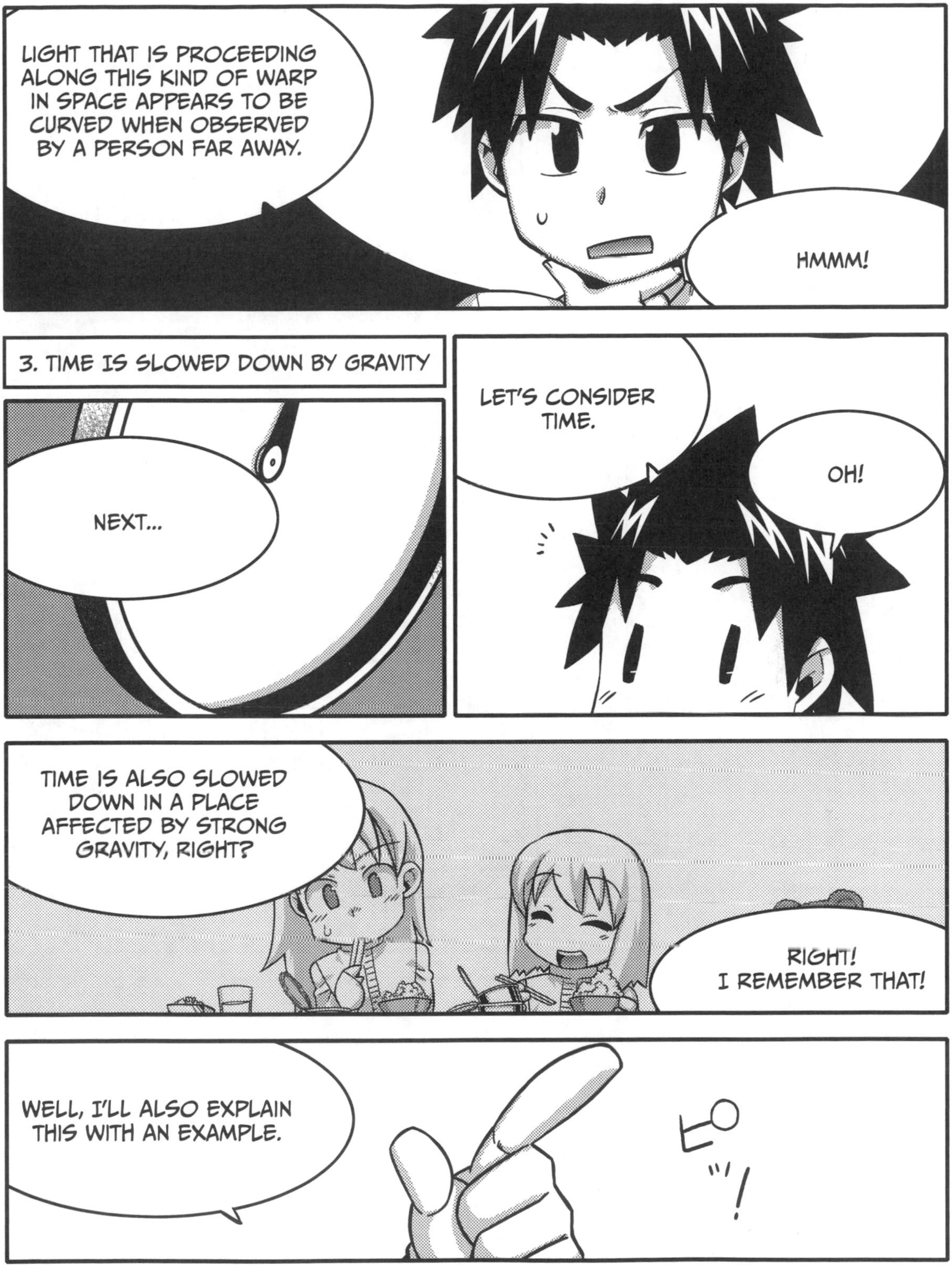
LIGHT THAT IS PROCEEDING ALONG THIS KIND OF WARP IN SPACE APPEARS TO BE CURVED WHEN OBSERVED BY A PERSON FAR AWAY.
HMMM!
3. TIME IS SLOWED DOWN BY GRAVITY
NEXT...
LET'S CONSIDER TIME.
OH!
TIME IS ALSO SLOWED DOWN IN A PLACE AFFECTED BY STRONG GRAVITY, RIGHT?
RIGHT! I REMEMBER THAT!
WELL, I'LL ALSO EXPLAIN THIS WITH AN EXAMPLE.
ピッ!

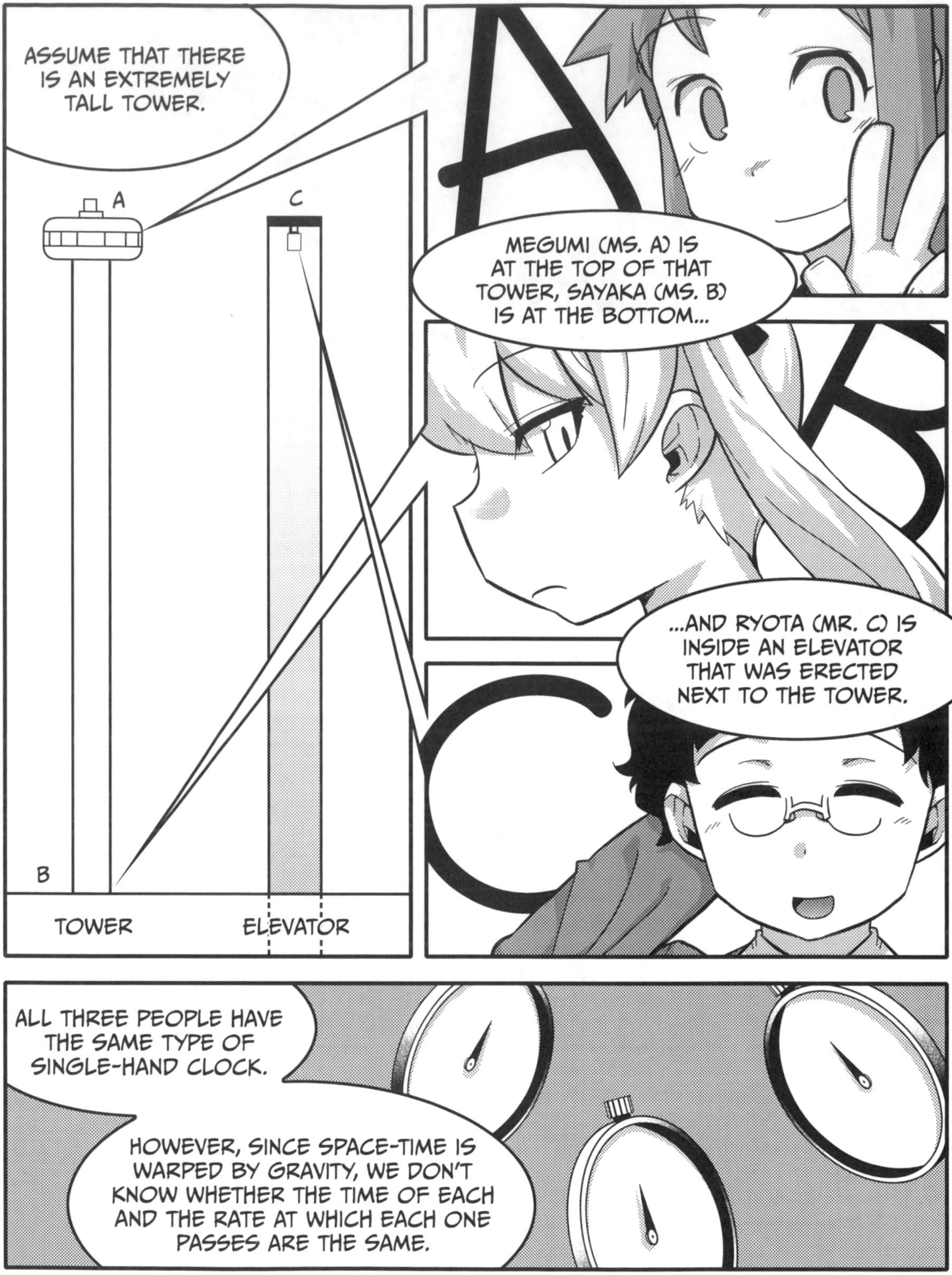
ASSUME THAT THERE IS AN EXTREMELY TALL TOWER.
A
C
B
TOWER
ELEVATOR
MEGUMI (MS. A) IS AT THE TOP OF THAT TOWER, SAYAKA (MS. B) IS AT THE BOTTOM...
...AND RYOTA (MR. C) IS INSIDE AN ELEVATOR THAT WAS ERECTED NEXT TO THE TOWER.
ALL THREE PEOPLE HAVE THE SAME TYPE OF SINGLE-HAND CLOCK.
HOWEVER, SINCE SPACE-TIME IS WARPED BY GRAVITY, WE DON'T KNOW WHETHER THE TIME OF EACH AND THE RATE AT WHICH EACH ONE PASSES ARE THE SAME.

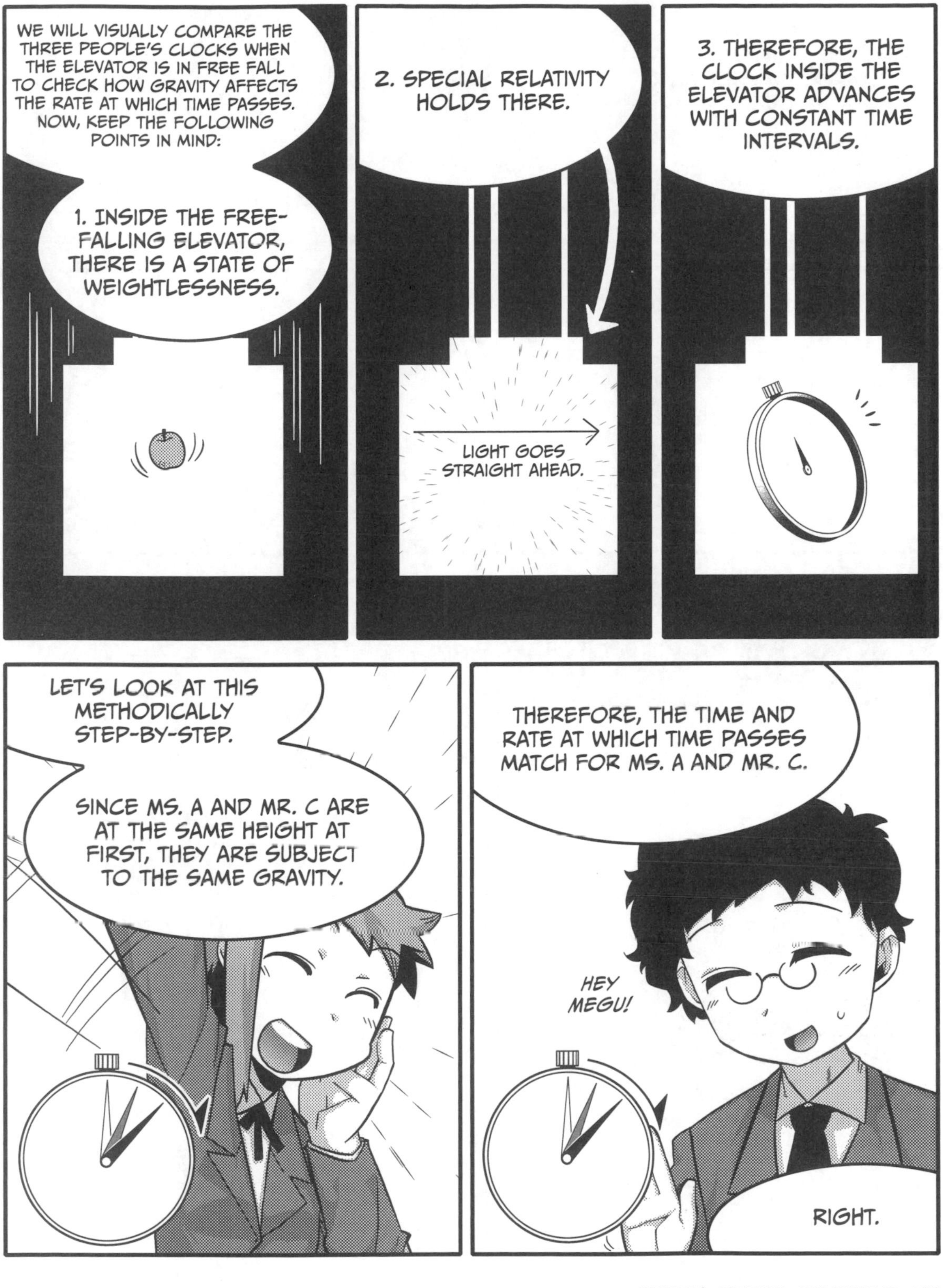
WE WILL VISUALLY COMPARE THE THREE PEOPLE'S CLOCKS WHEN THE ELEVATOR IS IN FREE FALL TO CHECK HOW GRAVITY AFFECTS THE RATE AT WHICH TIME PASSES. NOW, KEEP THE FOLLOWING POINTS IN MIND:
1. INSIDE THE FREE-FALLING ELEVATOR, THERE IS A STATE OF WEIGHTLESSNESS.
2. SPECIAL RELATIVITY HOLDS THERE.
LIGHT GOES STRAIGHT AHEAD.
3. THEREFORE, THE CLOCK INSIDE THE ELEVATOR ADVANCES WITH CONSTANT TIME INTERVALS.
LET'S LOOK AT THIS METHODICALLY STEP-BY-STEP.
SINCE MS. A AND MR. C ARE AT THE SAME HEIGHT AT FIRST, THEY ARE SUBJECT TO THE SAME GRAVITY.
THEREFORE, THE TIME AND RATE AT WHICH TIME PASSES MATCH FOR MS. A AND MR. C.
HEY MEGU!
RIGHT.

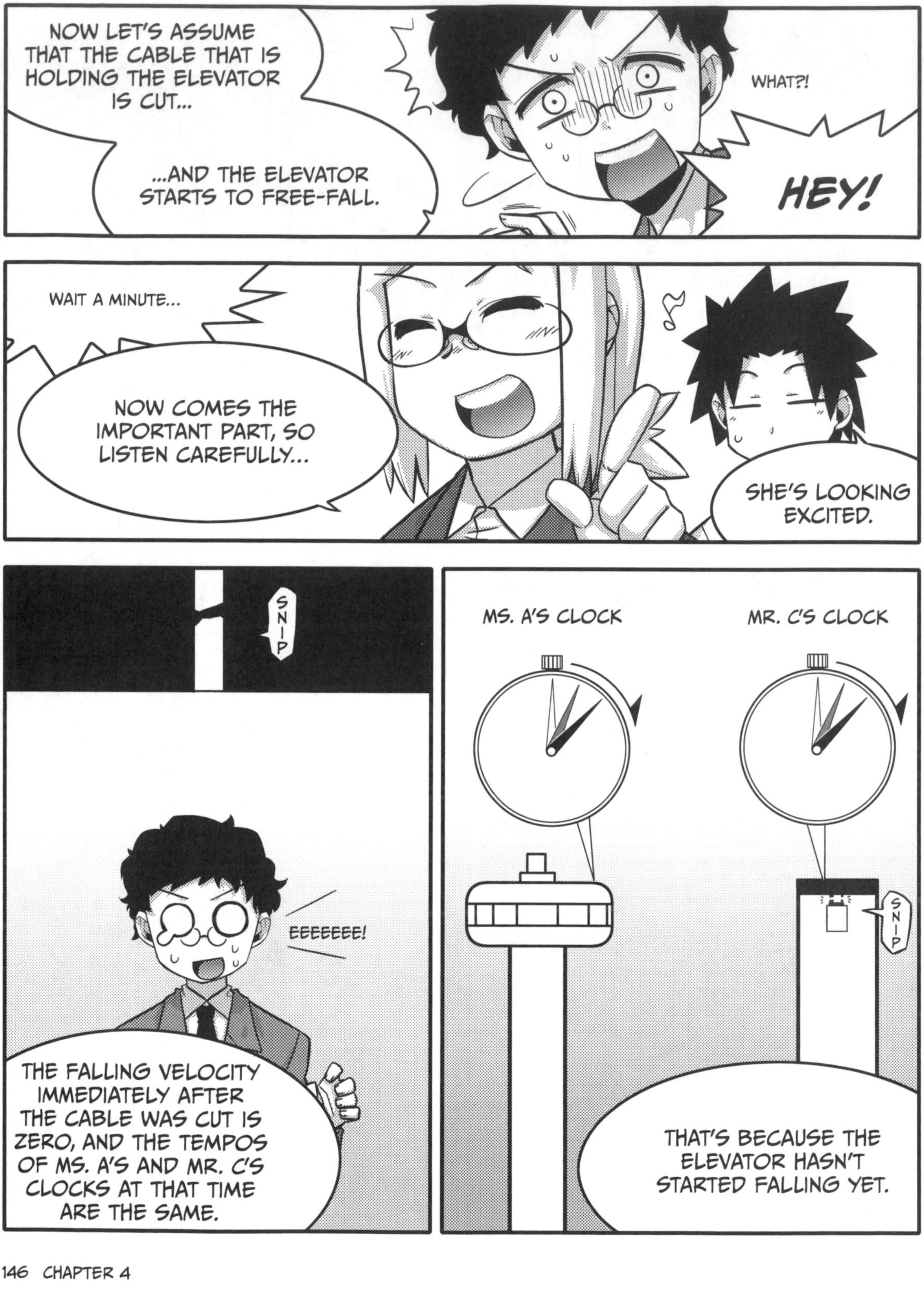
NOW LET'S ASSUME THAT THE CABLE THAT IS HOLDING THE ELEVATOR IS CUT...
...AND THE ELEVATOR STARTS TO FREE-FALL.
WHAT?!
HEY!
WAIT A MINUTE...
NOW COMES THE IMPORTANT PART, SO LISTEN CAREFULLY...
SHE'S LOOKING EXCITED.
SNIP
EEEEEEE!
THE FALLING VELOCITY IMMEDIATELY AFTER THE CABLE WAS CUT IS ZERO, AND THE TEMPOS OF MS. A'S AND MR. C'S CLOCKS AT THAT TIME ARE THE SAME.
MS. A'S CLOCK
MR. C'S CLOCK
SNIP
THAT'S BECAUSE THE ELEVATOR HASN'T STARTED FALLING YET.

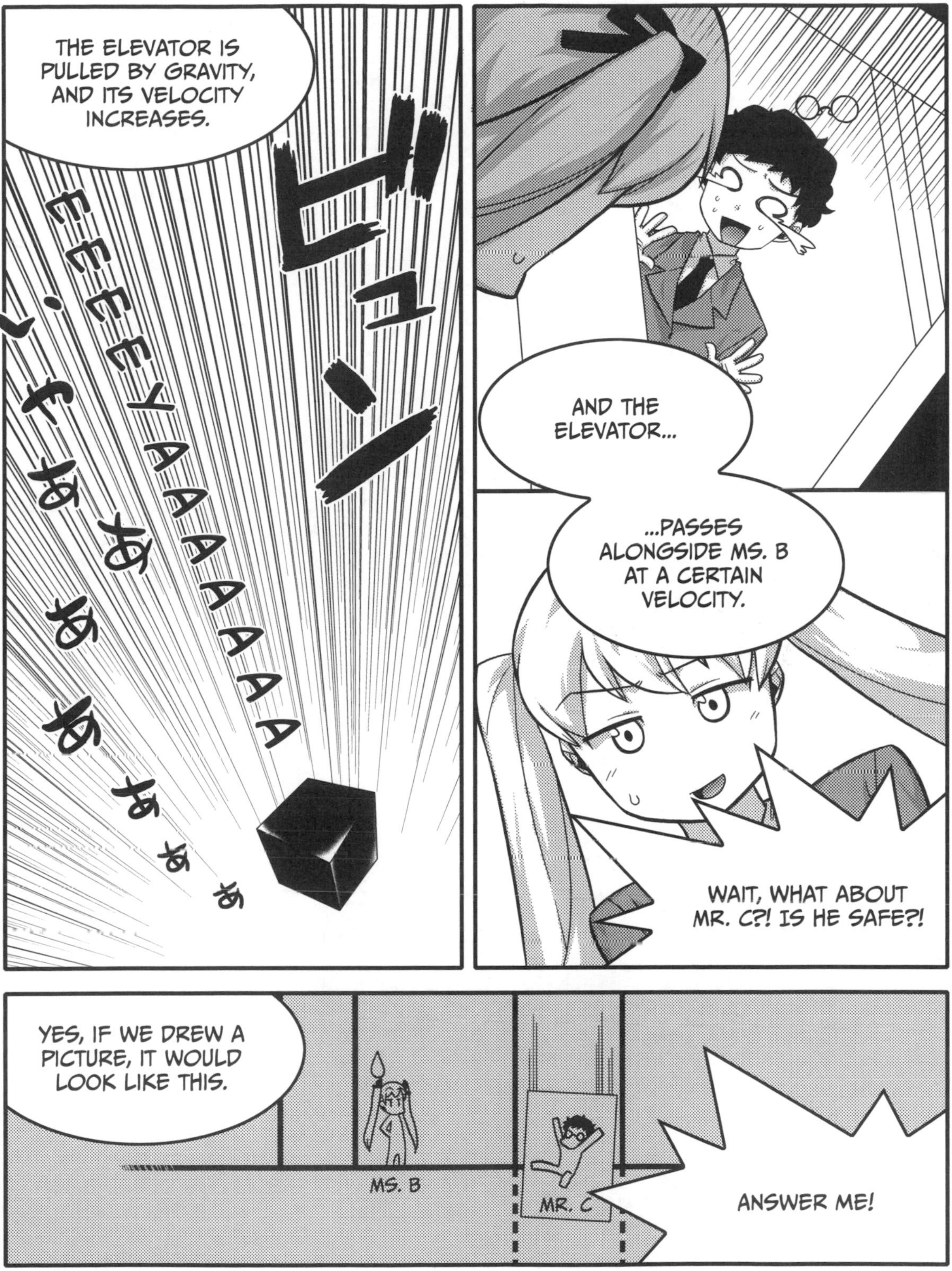
THE ELEVATOR IS PULLED BY GRAVITY, AND ITS VELOCITY INCREASES.
ビュン
EEEEYAAAAAAAA
いやあああああああ
AND THE ELEVATOR...
...PASSES ALONGSIDE MS. B AT A CERTAIN VELOCITY.
WAIT, WHAT ABOUT MR. C?! IS HE SAFE?!
YES, IF WE DREW A PICTURE, IT WOULD LOOK LIKE THIS.
MS. B
MR. C
ANSWER ME!

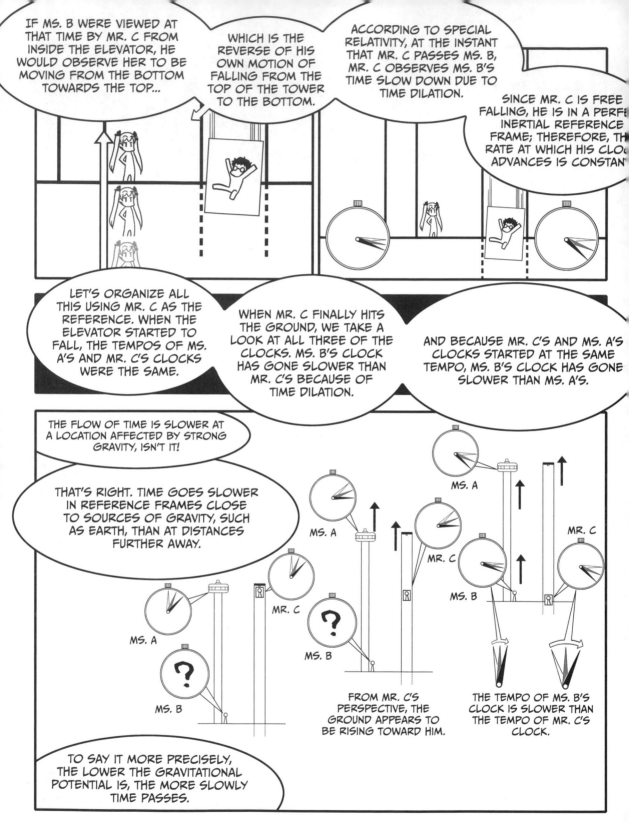

IF MS. B WERE VIEWED AT THAT TIME BY MR. C FROM INSIDE THE ELEVATOR, HE WOULD OBSERVE HER TO BE MOVING FROM THE BOTTOM TOWARDS THE TOP...

WHICH IS THE REVERSE OF HIS OWN MOTION OF FALLING FROM THE TOP OF THE TOWER TO THE BOTTOM.

ACCORDING TO SPECIAL RELATIVITY, AT THE INSTANT THAT MR. C PASSES MS. B, MR. C OBSERVES MS. B'S TIME SLOW DOWN DUE TO TIME DILATION.

SINCE MR. C IS FREE FALLING, HE IS IN A PERFE INERTIAL REFERENCE FRAME; THEREFORE, TH RATE AT WHICH HIS CLO ADVANCES IS CONSTAN

LET'S ORGANIZE ALL THIS USING MR. C AS THE REFERENCE. WHEN THE ELEVATOR STARTED TO FALL, THE TEMPOS OF MS. A'S AND MR. C'S CLOCKS WERE THE SAME.

WHEN MR. C FINALLY HITS THE GROUND, WE TAKE A LOOK AT ALL THREE OF THE CLOCKS. MS. B'S CLOCK HAS GONE SLOWER THAN MR. C'S BECAUSE OF TIME DILATION.

AND BECAUSE MR. C'S AND MS. A'S CLOCKS STARTED AT THE SAME TEMPO, MS. B'S CLOCK HAS GONE SLOWER THAN MS. A'S.

THE FLOW OF TIME IS SLOWER AT A LOCATION AFFECTED BY STRONG GRAVITY, ISN'T IT!

THAT'S RIGHT. TIME GOES SLOWER IN REFERENCE FRAMES CLOSE TO SOURCES OF GRAVITY, SUCH AS EARTH, THAN AT DISTANCES FURTHER AWAY.

MS. A

MS. B

MR. C

FROM MR. C'S PERSPECTIVE, THE GROUND APPEARS TO BE RISING TOWARD HIM.

MS. A

MR. C

MS. B

THE TEMPO OF MS. B'S CLOCK IS SLOWER THAN THE TEMPO OF MR. C'S CLOCK.

MS. A

MR. C

MS. B

TO SAY IT MORE PRECISELY, THE LOWER THE GRAVITATIONAL POTENTIAL IS, THE MORE SLOWLY TIME PASSES.

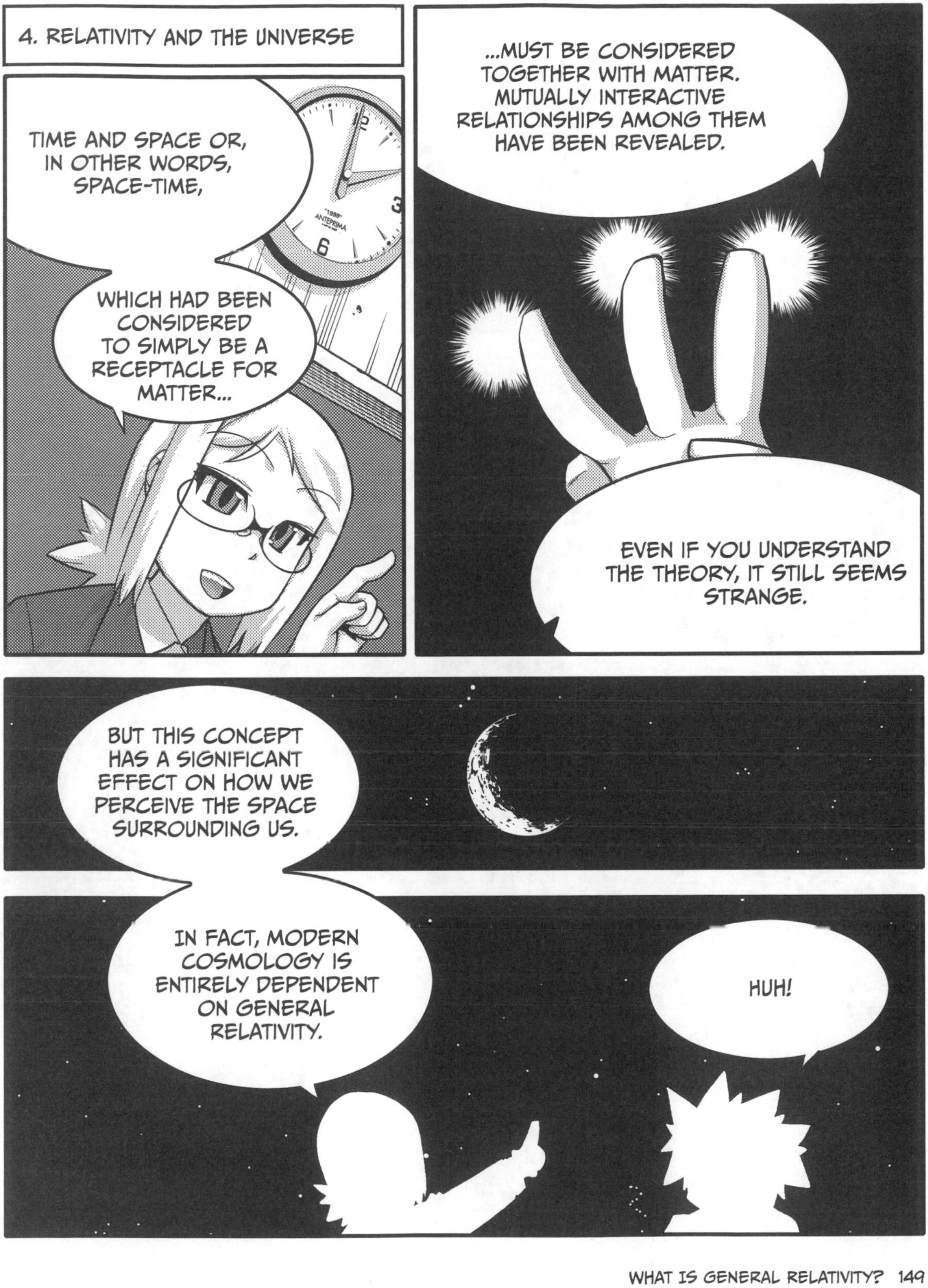
4. RELATIVITY AND THE UNIVERSE
TIME AND SPACE OR, IN OTHER WORDS, SPACE-TIME,
WHICH HAD BEEN CONSIDERED TO SIMPLY BE A RECEPTACLE FOR MATTER...
...MUST BE CONSIDERED TOGETHER WITH MATTER. MUTUALLY INTERACTIVE RELATIONSHIPS AMONG THEM HAVE BEEN REVEALED.
EVEN IF YOU UNDERSTAND THE THEORY, IT STILL SEEMS STRANGE.
BUT THIS CONCEPT HAS A SIGNIFICANT EFFECT ON HOW WE PERCEIVE THE SPACE SURROUNDING US.
IN FACT, MODERN COSMOLOGY IS ENTIRELY DEPENDENT ON GENERAL RELATIVITY.
HUH!

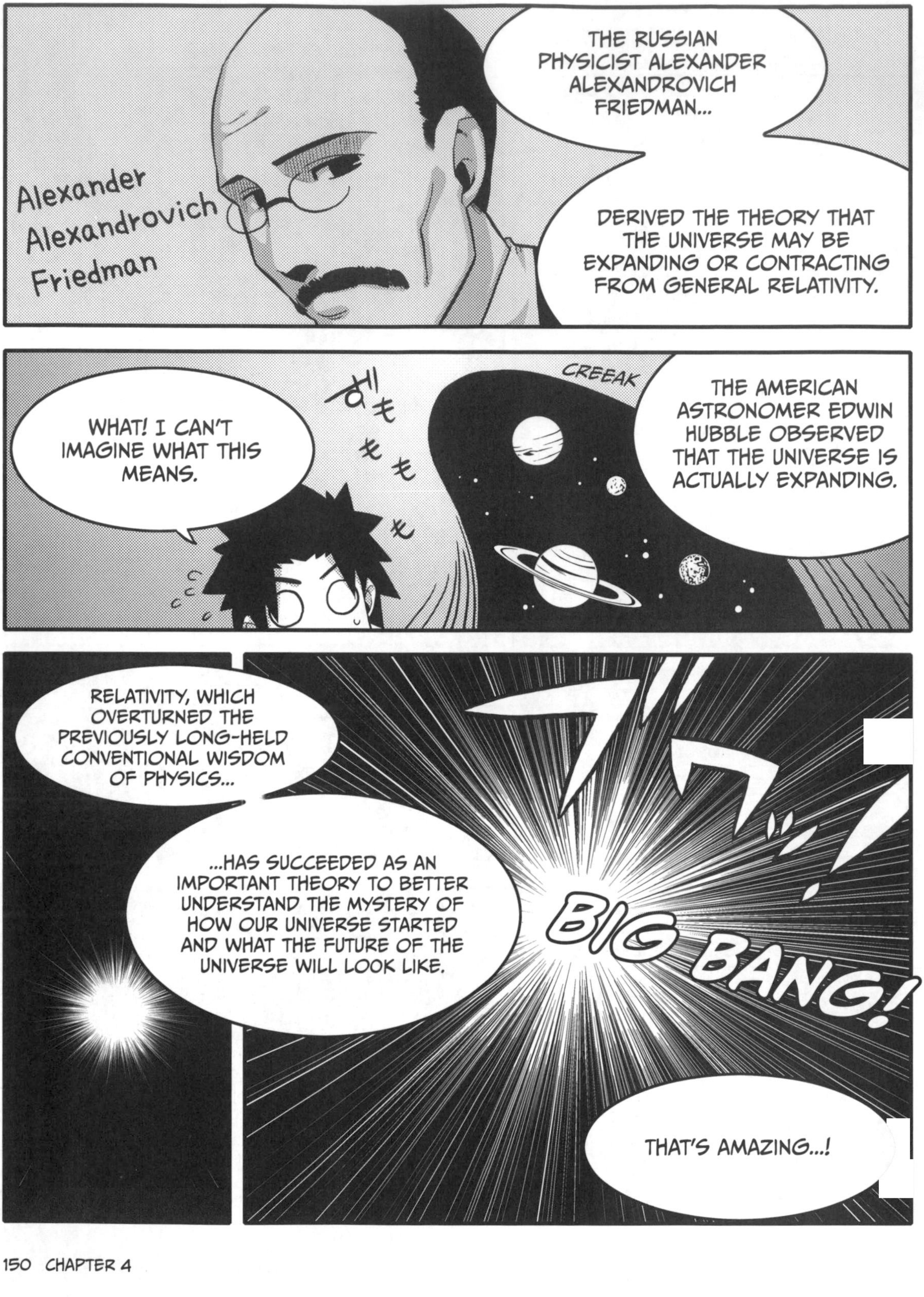
Alexander Alexandrovich Friedman
THE RUSSIAN PHYSICIST ALEXANDER ALEXANDROVICH FRIEDMAN...
DERIVED THE THEORY THAT THE UNIVERSE MAY BE EXPANDING OR CONTRACTING FROM GENERAL RELATIVITY.
WHAT! I CAN'T IMAGINE WHAT THIS MEANS.
CREEAK
THE AMERICAN ASTRONOMER EDWIN HUBBLE OBSERVED THAT THE UNIVERSE IS ACTUALLY EXPANDING.
RELATIVITY, WHICH OVERTURNED THE PREVIOUSLY LONG-HELD CONVENTIONAL WISDOM OF PHYSICS...
...HAS SUCCEEDED AS AN IMPORTANT THEORY TO BETTER UNDERSTAND THE MYSTERY OF HOW OUR UNIVERSE STARTED AND WHAT THE FUTURE OF THE UNIVERSE WILL LOOK LIKE.
BIG BANG!
THAT'S AMAZING...!

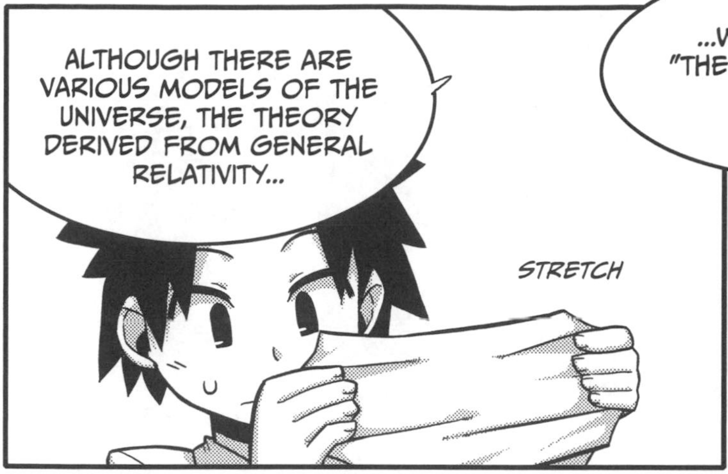
ALTHOUGH THERE ARE VARIOUS MODELS OF THE UNIVERSE, THE THEORY DERIVED FROM GENERAL RELATIVITY...
STRETCH
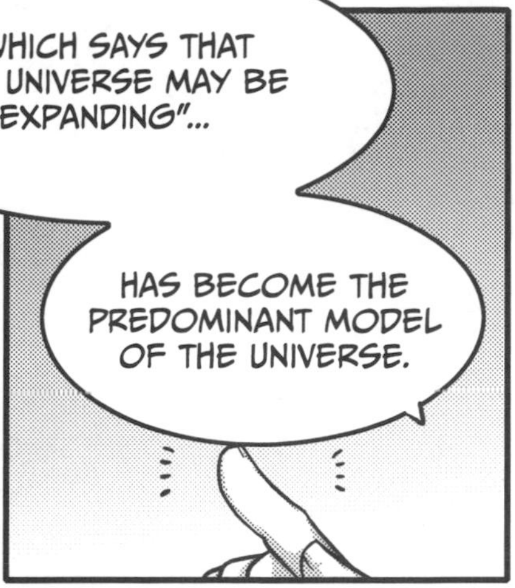
...WHICH SAYS THAT "THE UNIVERSE MAY BE EXPANDING"...
HAS BECOME THE PREDOMINANT MODEL OF THE UNIVERSE.
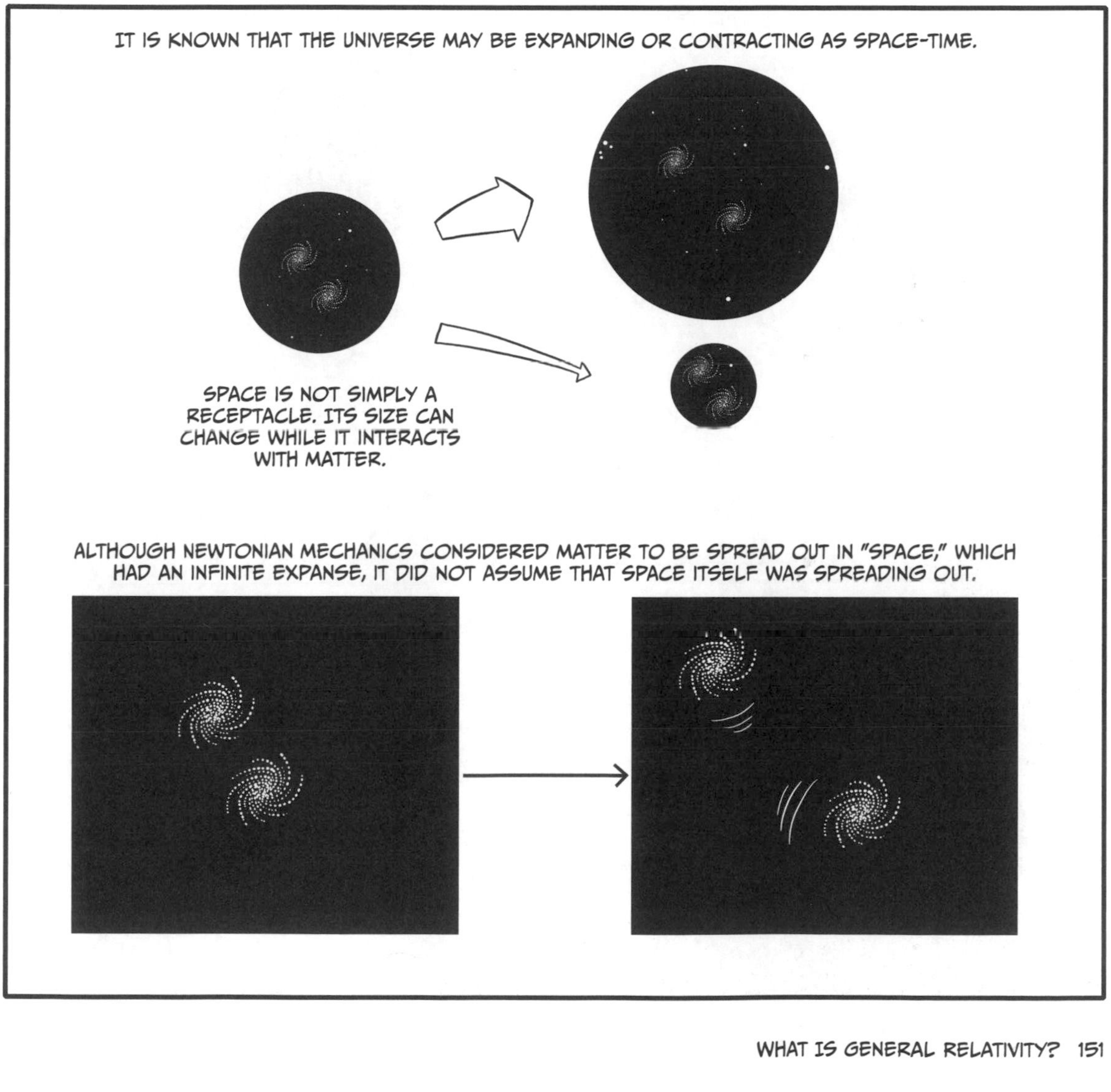
IT IS KNOWN THAT THE UNIVERSE MAY BE EXPANDING OR CONTRACTING AS SPACE-TIME.
SPACE IS NOT SIMPLY A RECEPTACLE. ITS SIZE CAN CHANGE WHILE IT INTERACTS WITH MATTER.
ALTHOUGH NEWTONIAN MECHANICS CONSIDERED MATTER TO BE SPREAD OUT IN "SPACE," WHICH HAD AN INFINITE EXPANSE, IT DID NOT ASSUME THAT SPACE ITSELF WAS SPREADING OUT.

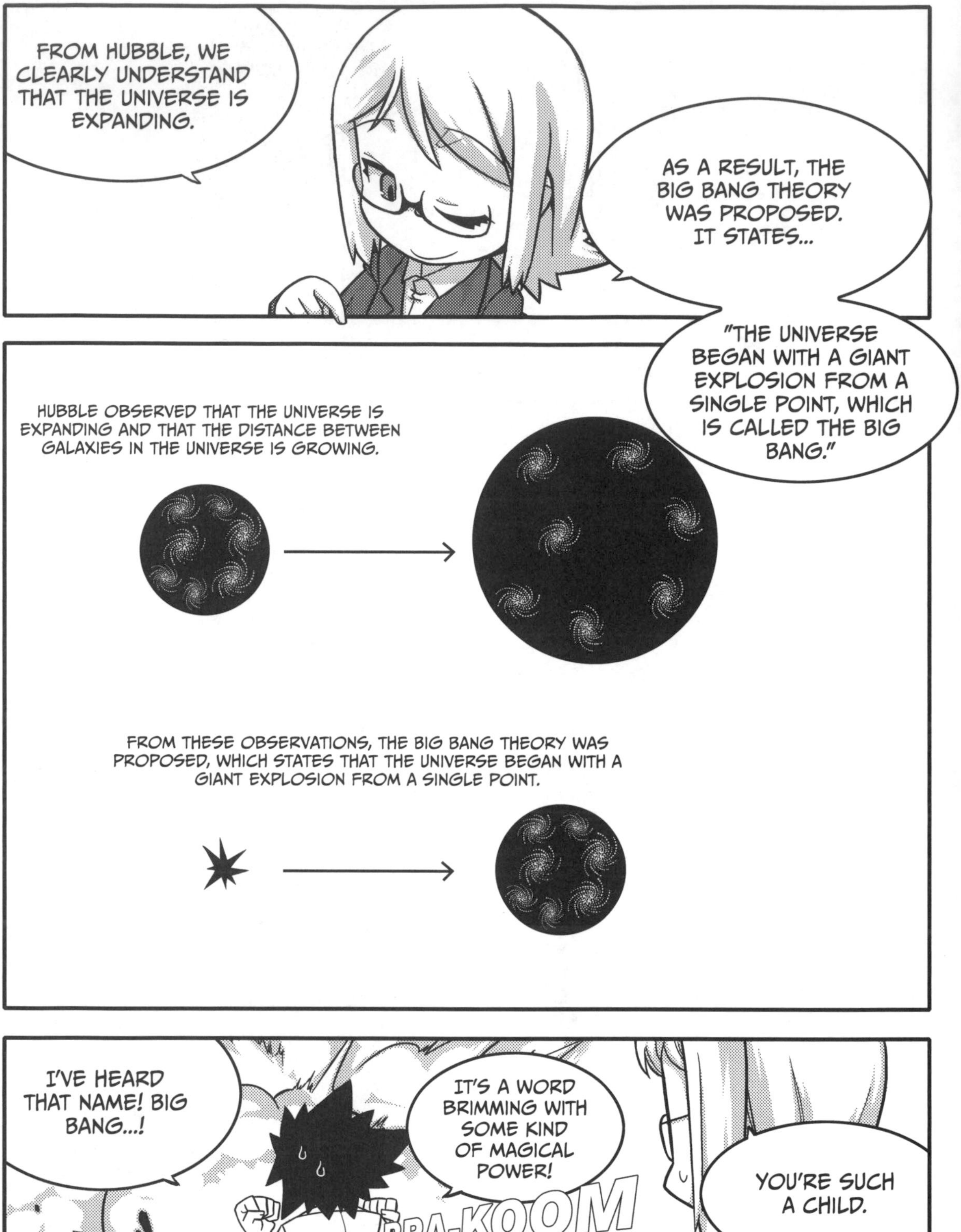

FROM HUBBLE, WE CLEARLY UNDERSTAND THAT THE UNIVERSE IS EXPANDING.

AS A RESULT, THE BIG BANG THEORY WAS PROPOSED. IT STATES...

"THE UNIVERSE BEGAN WITH A GIANT EXPLOSION FROM A SINGLE POINT, WHICH IS CALLED THE BIG BANG."

HUBBLE OBSERVED THAT THE UNIVERSE IS EXPANDING AND THAT THE DISTANCE BETWEEN GALAXIES IN THE UNIVERSE IS GROWING.

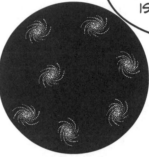

FROM THESE OBSERVATIONS, THE BIG BANG THEORY WAS PROPOSED, WHICH STATES THAT THE UNIVERSE BEGAN WITH A GIANT EXPLOSION FROM A SINGLE POINT.

I'VE HEARD THAT NAME! BIG BANG...!

IT'S A WORD BRIMMING WITH SOME KIND OF MAGICAL POWER!

BRA-KOOM

YOU'RE SUCH A CHILD.

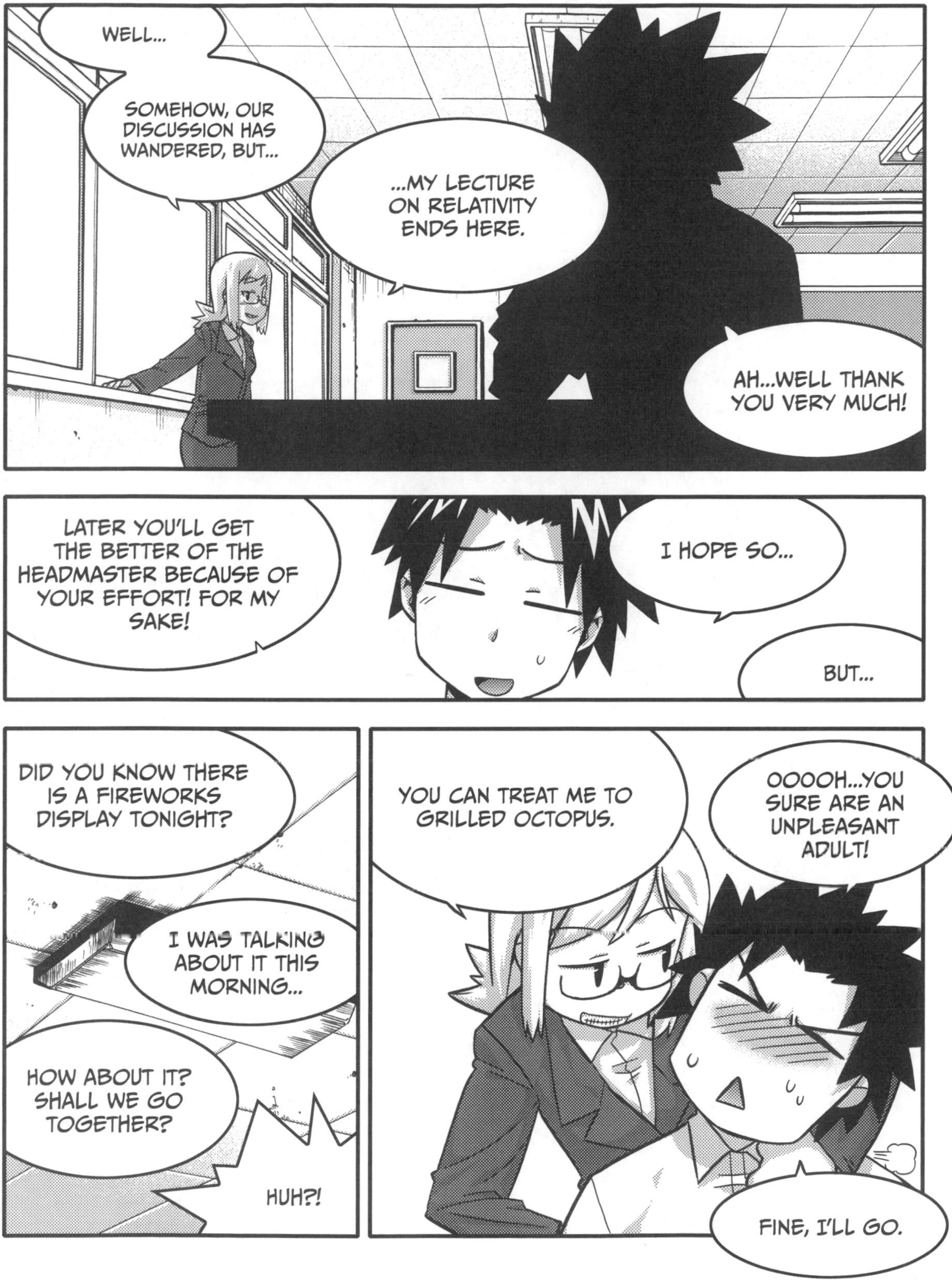
WELL...
SOMEHOW, OUR DISCUSSION HAS WANDERED, BUT...
...MY LECTURE ON RELATIVITY ENDS HERE.
AH...WELL THANK YOU VERY MUCH!
LATER YOU'LL GET THE BETTER OF THE HEADMASTER BECAUSE OF YOUR EFFORT! FOR MY SAKE!
I HOPE SO...
BUT...
DID YOU KNOW THERE IS A FIREWORKS DISPLAY TONIGHT?
I WAS TALKING ABOUT IT THIS MORNING...
HOW ABOUT IT? SHALL WE GO TOGETHER?
HUH?!
YOU CAN TREAT ME TO GRILLED OCTOPUS.
OOOOH...YOU SURE ARE AN UNPLEASANT ADULT!
FINE, I'LL GO.

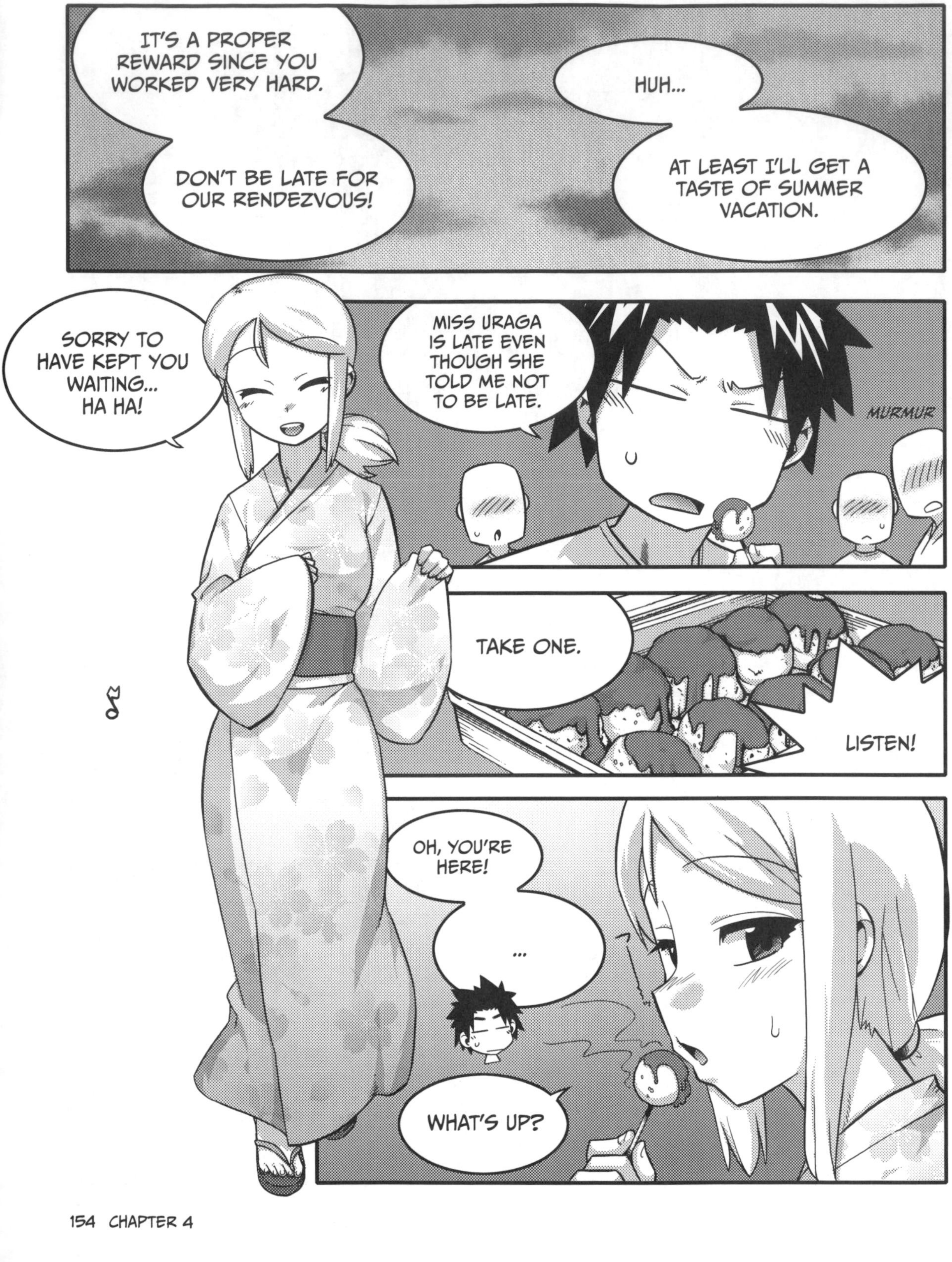
IT'S A PROPER REWARD SINCE YOU WORKED VERY HARD.
DON'T BE LATE FOR OUR RENDEZVOUS!
HUH...
AT LEAST I'LL GET A TASTE OF SUMMER VACATION.
SORRY TO HAVE KEPT YOU WAITING... HA HA!
MISS URAGA IS LATE EVEN THOUGH SHE TOLD ME NOT TO BE LATE.
MURMUR
TAKE ONE.
LISTEN!
OH, YOU'RE HERE!
...
WHAT'S UP?

YOU KNOW, MISS URAGA, YOU'RE PRETTY COOL WHEN YOU'RE OUTSIDE OF THE CLASSROOM.

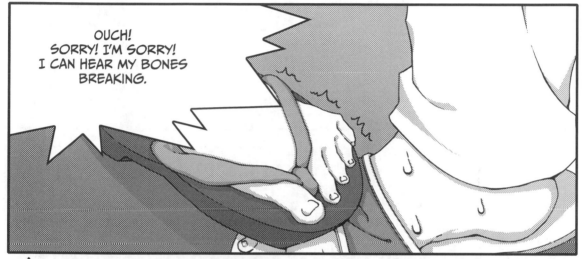

OUCH! SORRY! I'M SORRY! I CAN HEAR MY BONES BREAKING.

WOW... AREN'T THEY BEAUTIFUL?

IT'S NICE YOU ARE ENJOYING IT SO MUCH.

SILENCE

HUH?

IS IT OVER ALREADY?

OH! THERE GOES ANOTHER ONE!

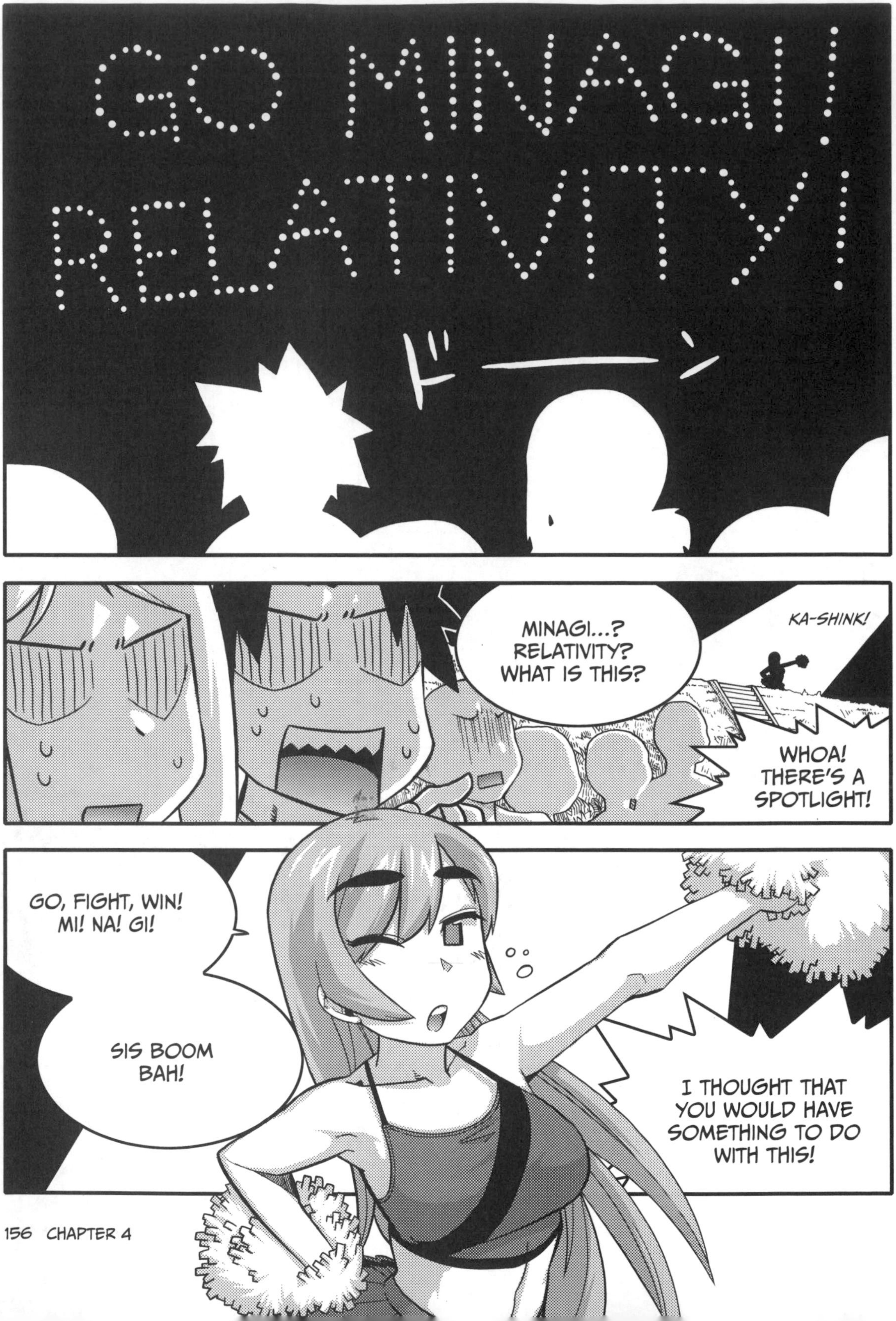
GO MINAGI! RELATIVITY!
ドーン
MINAGI...? RELATIVITY? WHAT IS THIS?
KA-SHINK!
WHOA! THERE'S A SPOTLIGHT!
GO, FIGHT, WIN! MI! NA! GI!
SIS BOOM BAH!
I THOUGHT THAT YOU WOULD HAVE SOMETHING TO DO WITH THIS!

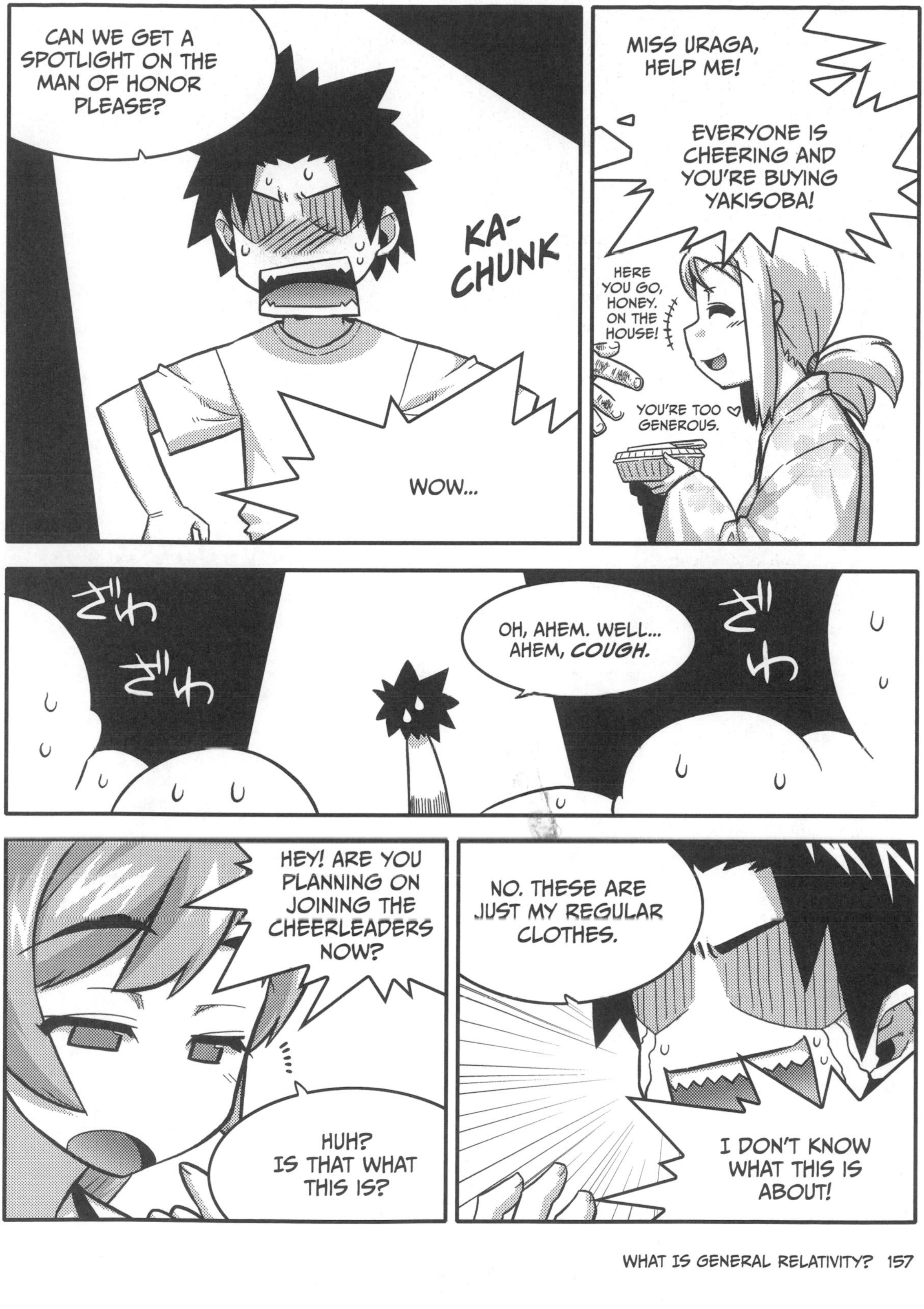
CAN WE GET A SPOTLIGHT ON THE MAN OF HONOR PLEASE?
KA-CHUNK
WOW...
MISS URAGA, HELP ME!
EVERYONE IS CHEERING AND YOU'RE BUYING YAKISOBA!
HERE YOU GO, HONEY. ON THE HOUSE!
YOU'RE TOO ♡ GENEROUS.
ざわ
ざわ
OH, AHEM. WELL... AHEM, *COUGH.*
ざわ
HEY! ARE YOU PLANNING ON JOINING THE CHEERLEADERS NOW?
HUH? IS THAT WHAT THIS IS?
NO. THESE ARE JUST MY REGULAR CLOTHES.
I DON'T KNOW WHAT THIS IS ABOUT!

THE SLOWING OF TIME IN GENERAL RELATIVITY

Let's use some equations to look at the "slowing of time" in general relativity based on the explanation in the manga.

As in the manga, we assume that Ms. A is at the top of a tall tower, Ms. B is at the bottom of the tower, and Mr. C is inside the elevator next to the tower, as shown in Figure 4-1.

We also assume that each of the three people has the same clock. However, since space-time is warped by gravity, we don't know whether the time of each and the rate at which each one passes (tempo) are the same.

Therefore, we will check the rate at which time passes due to gravity under the following three conditions:

1. Inside the free-falling elevator, there is a state of weightlessness.
2. Since special relativity holds there, the clock inside the elevator advances with constant time intervals.
3. Ms. A's clock at the top of the tower and Ms. B's clock at the bottom of the tower each advance with different constant time intervals.

In addition, we will use the following procedure to check the rate at which time passes due to gravity.

1. Align the rates at which time passes for Mr. C's clock inside the elevator and Ms. A's clock at the beginning of the descent.
2. Compare the rates at which time passes for Mr. C's clock inside the elevator and Ms. B's clock at the end of the descent.

At first, since Ms. A and Mr. C are at the same height, they are affected by the same gravity.

Let z denote the height direction at that location, and let ϕ_1 denote the gravitational potential. The gravitational potential is the quotient of the potential energy divided by the mass of an object. For example, the potential energy of gravity near the surface of the Earth is mgh, and the gravitational potential is gh.

Therefore, we will align Ms. A's and Mr. C's times and the rates at which time passes.

Let $\Delta\tau_1$ denote the time that passes at Ms. A's location, and let $\Delta\tau_2$ denote the time that passes at Ms. B's location.

Now let's assume that the cable that is holding the elevator is cut and the elevator starts to free-fall. Since the falling velocity immediately after the cable is cut (the velocity at which Ms. A is flying upward when viewed from Mr. C's perspective) is $v = 0$, the tempos of Ms. A's and Mr. C's clocks are the same.

❶ $\quad \Delta\tau_1 = \Delta\tau_3$

The elevator is pulled by gravity, and its velocity gradually increases. The elevator passes alongside Ms. B at a certain velocity (v).

If Ms. B were viewed at that time by Mr. C inside the elevator, he would observe her to be moving upward toward himself, which is the reverse of his own motion (falling from the top of the tower toward the bottom) viewed from his surroundings (see Figure 4-2).

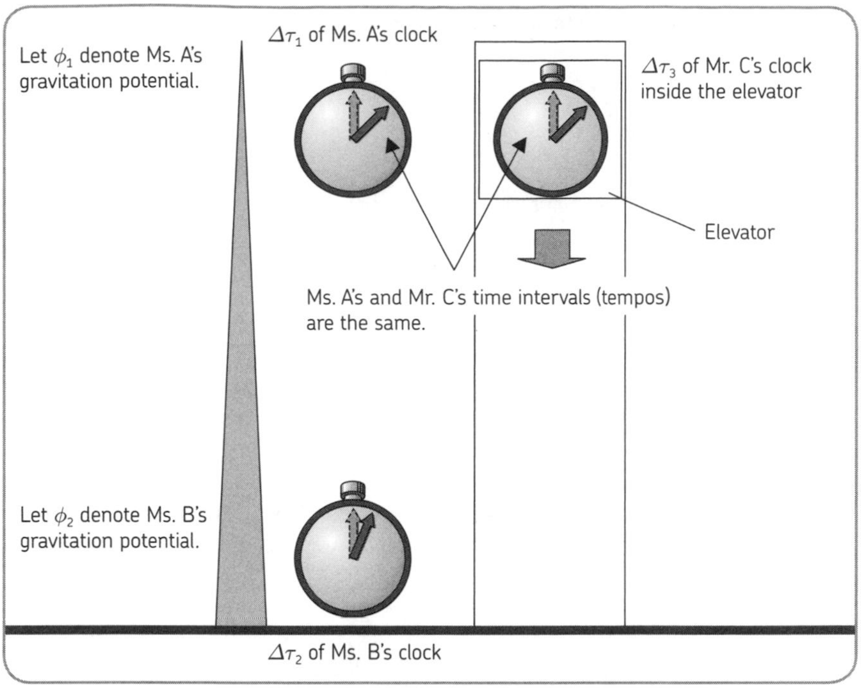
$\Delta\tau_1$ of Ms. A's clock
Let ϕ_1 denote Ms. A's gravitation potential.
$\Delta\tau_3$ of Mr. C's clock inside the elevator
Elevator
Ms. A's and Mr. C's time intervals (tempos) are the same.
Let ϕ_2 denote Ms. B's gravitation potential.
$\Delta\tau_2$ of Ms. B's clock

Figure 4-1: Aligning the rates at which time passes for Mr. C's clock inside the elevator and Ms. A's clock at the beginning of the descent

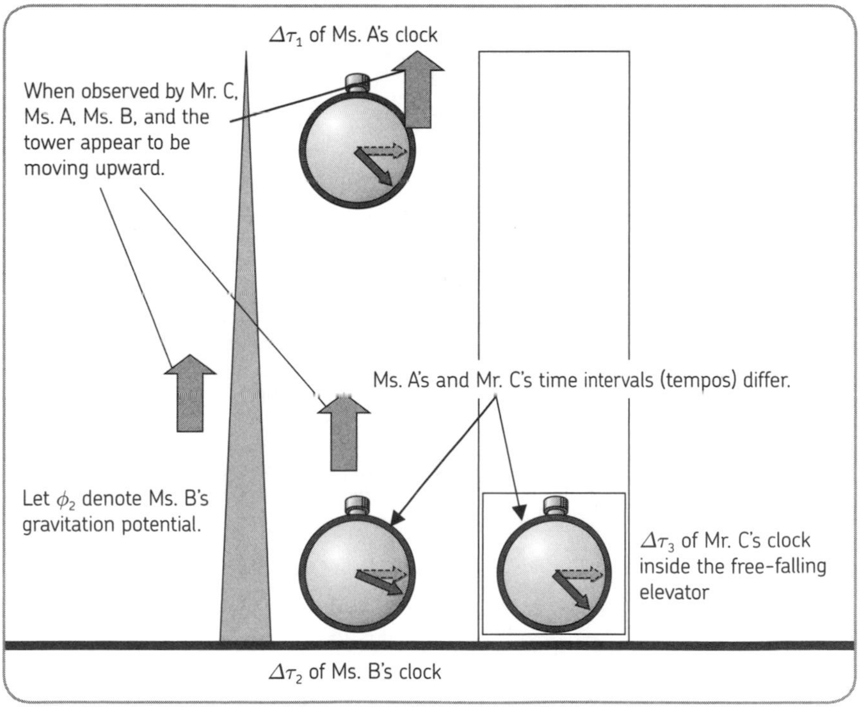
$\Delta\tau_1$ of Ms. A's clock
When observed by Mr. C, Ms. A, Ms. B, and the tower appear to be moving upward.
Ms. A's and Mr. C's time intervals (tempos) differ.
Let ϕ_2 denote Ms. B's gravitation potential.
$\Delta\tau_3$ of Mr. C's clock inside the free-falling elevator
$\Delta\tau_2$ of Ms. B's clock

Figure 4-2: Comparing the rates at which time passes for the clocks of Mr. C inside the elevator and Ms. B at the end of the descent

According to special relativity, at the instant that Mr. C passed Ms. B, we have

❷ $\qquad \Delta\tau_2 = \Delta\tau_3 \sqrt{1 - \left(\dfrac{v}{c}\right)^2}$,

and from equations ❶ and ❷, we eliminate $\Delta\tau_3$ to obtain

❸ $\qquad \dfrac{\Delta\tau_2}{\Delta\tau_1} = \sqrt{1 - \left(\dfrac{v}{c}\right)^2} < 1$.

This shows that the tempo of Ms. B's clock becomes slower than the tempo of Mr. C's clock. The rate at which time passes for Ms. B's clock, where the gravitational potential is low (ϕ_2, close to the source of gravity), is slower than the rate at which time passes for Ms. A's clock, where the gravitational potential is high (ϕ_1, far from the source of gravity).

In other words, the lower the gravitational potential is, the slower time will pass.

Let's assume that the velocity v is low and that we can use Newtonian mechanics (if we let $x = \dfrac{v}{c}$, then $x \ll 1$).

Therefore, when ϕ_1 denotes the gravitational potential at Ms. A's location and ϕ_2 denotes the gravitational potential at Ms. B's location, we have $\phi_1 > \phi_2$.

Due to the conservation of energy, at the instant before the elevator hits the ground, all of its potential energy has been converted to kinetic energy, and its velocity v is given by this expression:

$$(\phi_1 - \phi_2)m = \frac{1}{2}mv^2$$

We obtain the following:

❹ $\qquad \phi_1 - \phi_2 = \dfrac{1}{2}v^2$

When $x \ll 1$, we can use the following approximation formula:

$$(1+x)^\alpha \approx 1 + \alpha x$$

Since $x \ll 1$ because $x = \dfrac{v}{c}$, we have the following:

$$\sqrt{1 - \left(\frac{v}{c}\right)^2} = \left(1 - x^2\right)^{\frac{1}{2}} \approx 1 - \frac{1}{2}x^2 = 1 - \frac{1}{2}\left(\frac{v}{c}\right)^2$$

If we use this together with equation ❸, we obtain the following equation:

❺ $$\frac{\Delta\tau_2}{\Delta\tau_1} = \sqrt{1-\left(\frac{v}{c}\right)^2} \approx 1-\frac{1}{2}\left(\frac{v}{c}\right)^2$$

If we substitute $\frac{1}{2}v^2 = \phi_1 - \phi_2$ from equation ❹ into equation ❺, we obtain the following equation:

❻ $$\frac{\Delta\tau_2}{\Delta\tau_1} \approx 1-\frac{1}{2}\left(\frac{v}{c}\right)^2 - 1 - \frac{\phi_1-\phi_2}{c^2}$$

Also, if we rearrange the above equation slightly, since $\frac{\phi_1-\phi_2}{c^2} \approx 1-\frac{\Delta\tau_2}{\Delta\tau_1} = \frac{\Delta\tau_1-\Delta\tau_2}{\Delta\tau_1}$ holds, we have the following:

❼ $$\frac{\Delta\tau_1-\Delta\tau_2}{\Delta\tau_1} \approx \frac{\phi_1-\phi_2}{c^2}$$

In other words, gravitational potential is related to the slowing of time (time dilation), as shown in equation ❼.

As shown in Figure 4-3, the difference in gravitational potential $\phi_1 - \phi_2$ between the ground and an object at a height h is given by the gravitational potential gh.

$\Delta\tau_1$ of Ms. A's clock
Gravitational potential ϕ_1
Height h
The two time intervals (tempos) differ.
$\Delta\tau_2 \neq \Delta\tau_3$
$\Delta\tau_1 = \Delta\tau_3 > \Delta\tau_2$
$\Delta\tau_3$ of Mr. C's clock inside the free-falling elevator
Gravitational potential ϕ_2
$\Delta\tau_2$ of Ms. B's clock

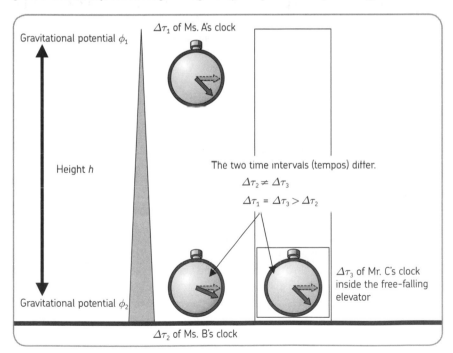

Figure 4-3: State of relatively weak gravity on the ground

If we let $\phi_2 = 0$, h denote the height up to ϕ_1, and g denote the gravitational acceleration near the ground and then substitute $\phi_1 = gh$ and $\phi_2 = 0$ into equation ❼, we obtain the following:

$$\frac{\Delta\tau_1 - \Delta\tau_2}{\Delta\tau_1} \approx \frac{\phi_1 - \phi_2}{c^2} = \frac{gh - 0}{c^2} = \frac{gh}{c^2}$$

The clock at the higher altitude measures a time $\Delta\tau_1$ that is ahead of the clock below, which reads $\Delta\tau_2$. We know this to be true, because in the equation above, gh/c^2 and $\Delta\tau_1$ are both greater than zero, and therefore $\Delta\tau_1 - \Delta\tau_2 > 0$; hence, $\Delta\tau_1 > \Delta\tau_2$.

THE TRUE NATURE OF GRAVITY IN GENERAL RELATIVITY

Space-time surrounding the existence of matter is warped, as explained in the manga. It is apparent that this warping of space-time has the same effect as gravity in attracting surrounding matter.

Einstein unified these effects in a set of equations called the *Einstein field equations*. The Einstein field equations showed that time and space (space-time), which were previously thought to exist as a framework for measuring the motion of matter, were fundamentally connected with matter itself.

PHENOMENA DISCOVERED FROM GENERAL RELATIVITY

This section introduces the following phenomena, which were discovered from general relativity:

*   Gravitational lensing
*   Anomalous perihelion precession of Mercury
*   Black holes

BENDING OF LIGHT (GRAVITATIONAL LENSING) NEAR A LARGE MASS (SUCH AS THE SUN)

Gravitational lensing is the phenomenon that when light passes in the vicinity of the Sun, the path of that light bends.

Space is bent in the vicinity of the Sun because of the large mass of the Sun, as shown in Figure 4-4. Since light advances along that curvature, the light from a distant star bends, and the direction of the star is observed with a slight shift. This effect was verified during a total solar eclipse. It is noted as the first proof discovered of general relativity.

Also, when light is coming from a distant galaxy, as shown in Figure 4-5, if a massive object (such as a galaxy) lies at an intermediate point, it will bend the light from the distant galaxy as though there were a condensing lens at that intermediate point. This bend may make the distant galaxy seem distorted. Many instances of this effect have been observed. This is another instance of gravitational lensing.

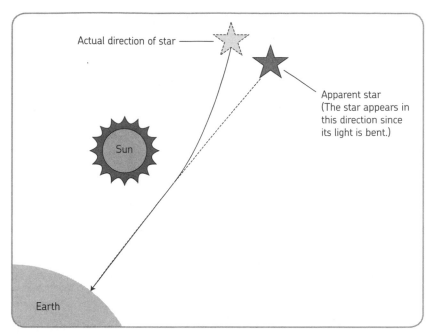

Actual direction of star

Apparent star
(The star appears in
this direction since
its light is bent.)

Sun

Earth

Figure 4-4: Bending of light near a large mass

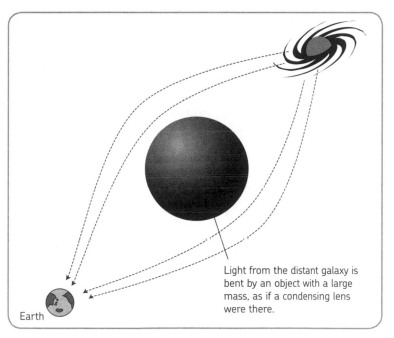

Light from the distant galaxy is
bent by an object with a large
mass, as if a condensing lens
were there.

Earth

Figure 4-5: Gravitational lensing

ANOMALOUS PERIHELION PRECESSION OF MERCURY

The *perihelion* is the point in the orbit of a planet that is closest to the Sun, as shown in Figure 4-6. We know that the perihelion of Mercury moves by approximately 574 arc seconds per century. Note that the "seconds" mentioned here are angular units rather than units of time. A minute of an arc is 1/60 of 1 degree, and a second is 1/60 of that. In other words, an arc second is 1/3600 of a degree. If the shift revolves by approximately 547 arc seconds per century, it is a shift of only approximately 0.16 degree per century.

Mercury
Nearest point to the Sun
(perihelion)
Sun

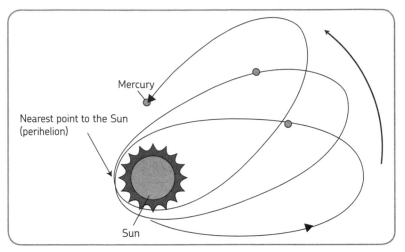

Figure 4-6: The anomalous shift in the perihelion of Mercury

Various causes of this shift, such as the effects of gravity of other planets, were investigated using Newtonian mechanics. However, none of these investigations were able to account for 43 arc seconds of perihelion movement. But when general relativity was used to check the shift in the perihelion of Mercury by calculating the Sun's warping of spacetime, it was found to be shifted by exactly 43 arc seconds.

BLACK HOLES

A *black hole* is a condition in which mass is extremely concentrated and gravity becomes so strong that not even light can escape from it.

A supernova stellar explosion occurs at the end of the life of a star having a mass several times that of the Sun. This event forms a region in space in which mass is extremely concentrated and gravity becomes stronger. Gravity becomes so strong that even light may not be able to escape. That region is a black hole.

Since light cannot escape from it, a black hole cannot be directly observed. However, if other stars exist in the vicinity of the black hole, gas from those stars will stream toward the black hole and form an accretion disk, that is, a cloud of diffuse matter around the black hole. When gas streams into the black hole from that accretion disk, X-rays and gamma rays are emitted.

A black hole candidate was discovered in 1971 in the constellation Cygnus, and currently, supermassive black holes are believed to exist at the center of galactic systems.

GLOBAL POSITIONING SYSTEM AND RELATIVITY

The Global Positioning System (GPS) uses 24 satellites orbiting the Earth to determine position. Each satellite broadcasts a signal toward Earth that includes the radio broadcast time. A receiver on the ground (such as a car navigation system) receives those signals. The radio waves of the signals reach the receiver at the speed of light (approximately 300,000,000 m/s).

When the time the signal was received is compared with the broadcast time, and that time difference is multiplied by the speed of light, the distance to the satellite is known. In other words, if we assume that the distance between the satellite and receiver is 20,000 km, then the radio waves reach the receiver in 20,000,000 m ÷ 300,000,000 m/s = 0.067 seconds. That calculation is performed using radio waves from three satellites to accurately determine the position on the ground.

However, if there is an error in that time difference, an error will also occur in the calculation of the distance between the satellite and receiver. For example, if the satellite broadcast time is offset by 1 microsecond (10^{-6} second), the distance will be offset by 300 meters (300,000,000 m/s × 0.000001 s = 300 m).

A GPS satellite orbits the Earth at an altitude of 20,000 km and a velocity that causes it to make 1 revolution in approximately 12 hours. At that velocity, the effect of special relativity causes its time to slow by 7.1 microseconds per day. However, since it is located high above the surface of the Earth, the effect of general relativity causes its time to pass faster than time on the surface of the Earth by 46.3 microseconds per day (the effect of general relativity is represented by equation ❼ on page 161). As a result, the time broadcast from the GPS is slowed by 39.2 microseconds per day (see Figure 4-7). The design of the GPS system takes into consideration the effects of both special and general relativity with extreme precision.

Since it is moving at a high velocity, its time is slower than time on the surface of the Earth by 7.1 microseconds per day due to the effect of special relativity.
Earth
GPS
Since it is located high above the surface of the Earth, gravity is weaker and its time passes faster than time on the surface of the Earth by 46.3 microseconds per day due to the effect of general relativity.

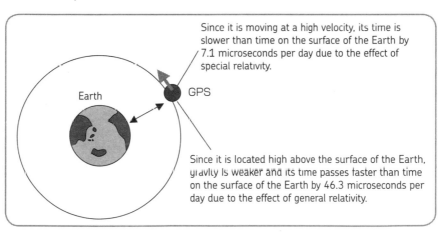

Figure 4-7: The Global Positioning System compensates for the effects of relativity.

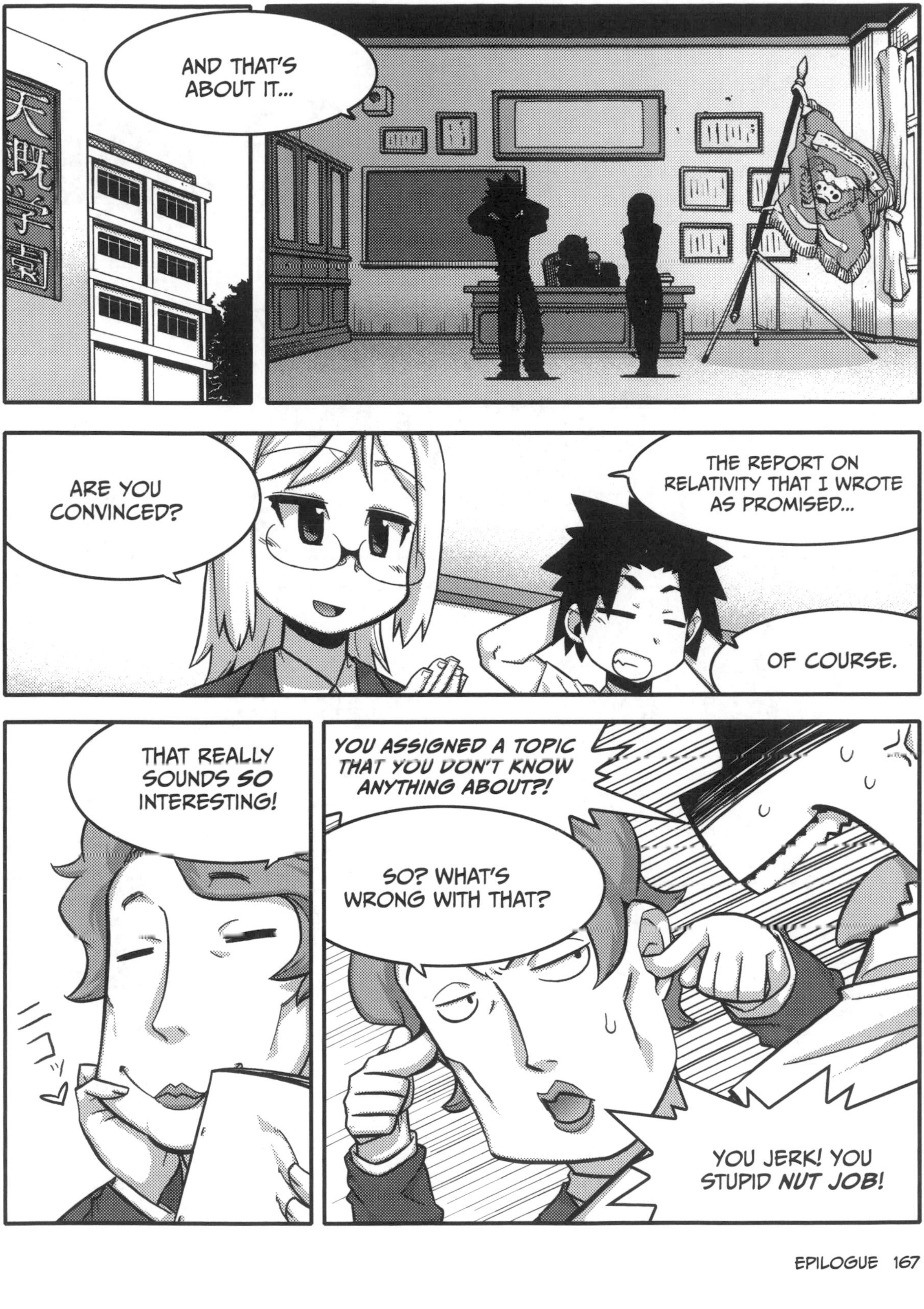
AND THAT'S ABOUT IT...
ARE YOU CONVINCED?
THE REPORT ON RELATIVITY THAT I WROTE AS PROMISED...
OF COURSE.
THAT REALLY SOUNDS *SO* INTERESTING!
ニコッ
YOU ASSIGNED A TOPIC THAT YOU DON'T KNOW ANYTHING ABOUT?!
SO? WHAT'S WRONG WITH THAT?
YOU JERK! YOU STUPID *NUT JOB!*

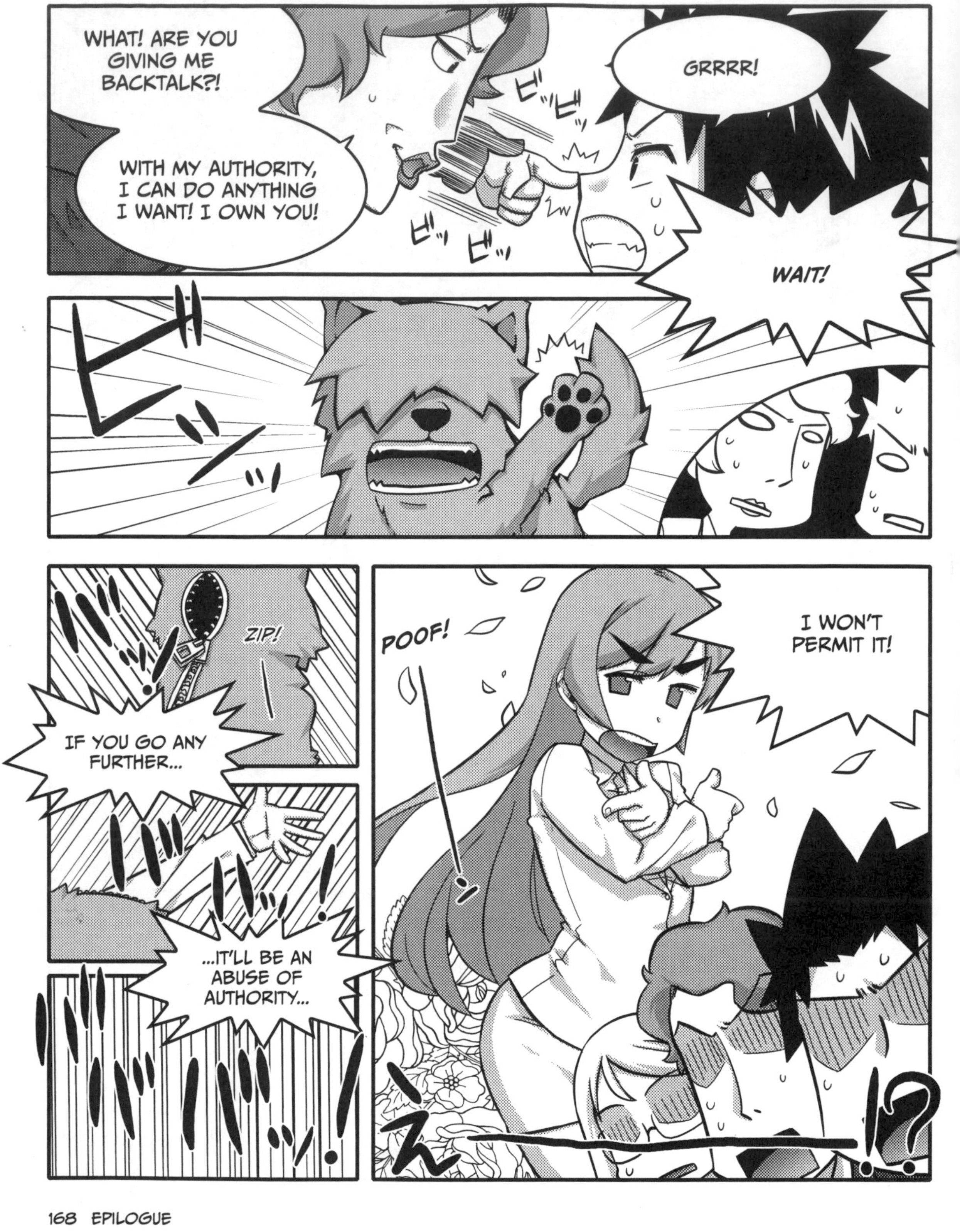
WHAT! ARE YOU GIVING ME BACKTALK?!
WITH MY AUTHORITY, I CAN DO ANYTHING I WANT! I OWN YOU!
GRRRR!
WAIT!
ZIP!
IF YOU GO ANY FURTHER...
...IT'LL BE AN ABUSE OF AUTHORITY...
POOF!
I WON'T PERMIT IT!

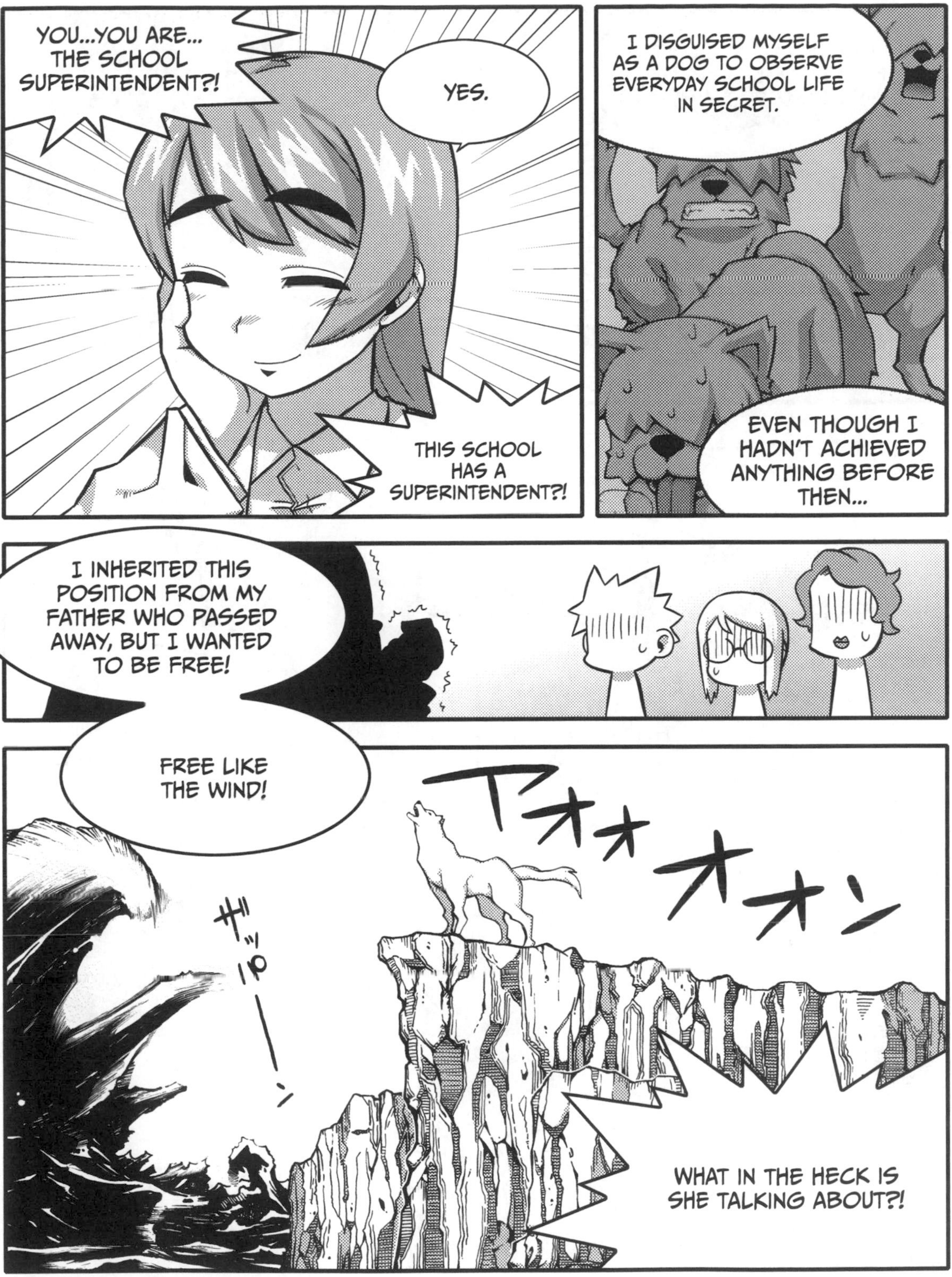
YOU...YOU ARE... THE SCHOOL SUPERINTENDENT?!
YES.
THIS SCHOOL HAS A SUPERINTENDENT?!
I DISGUISED MYSELF AS A DOG TO OBSERVE EVERYDAY SCHOOL LIFE IN SECRET.
EVEN THOUGH I HADN'T ACHIEVED ANYTHING BEFORE THEN...
I INHERITED THIS POSITION FROM MY FATHER WHO PASSED AWAY, BUT I WANTED TO BE FREE!
FREE LIKE THE WIND!
アオオオン
ザッパーン
WHAT IN THE HECK IS SHE TALKING ABOUT?!

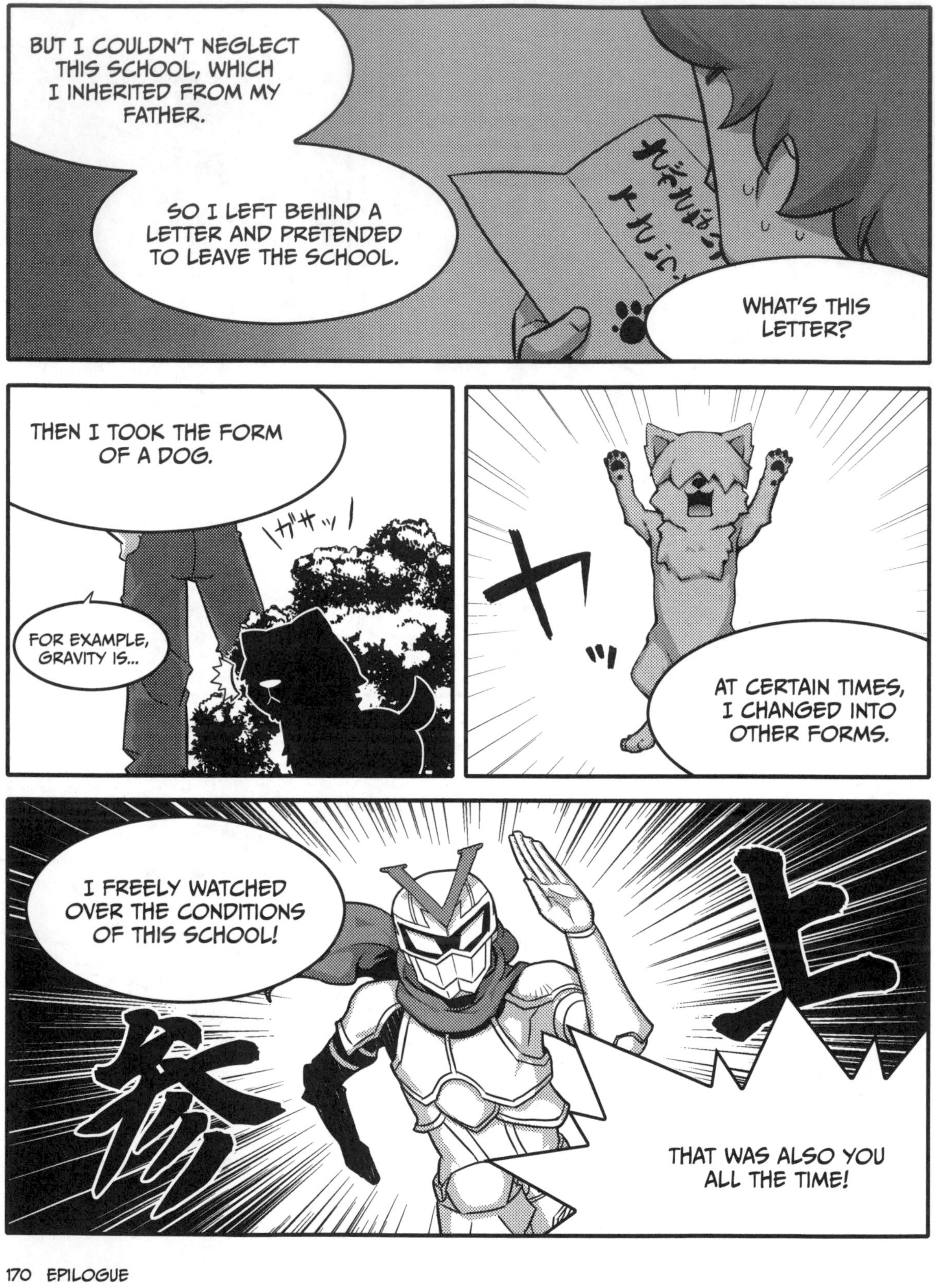
BUT I COULDN'T NEGLECT THIS SCHOOL, WHICH I INHERITED FROM MY FATHER.
SO I LEFT BEHIND A LETTER AND PRETENDED TO LEAVE THE SCHOOL.
WHAT'S THIS LETTER?
THEN I TOOK THE FORM OF A DOG.
ガサッ
FOR EXAMPLE, GRAVITY IS...
ヤッ
AT CERTAIN TIMES, I CHANGED INTO OTHER FORMS.
I FREELY WATCHED OVER THE CONDITIONS OF THIS SCHOOL!
参上
THAT WAS ALSO YOU ALL THE TIME!

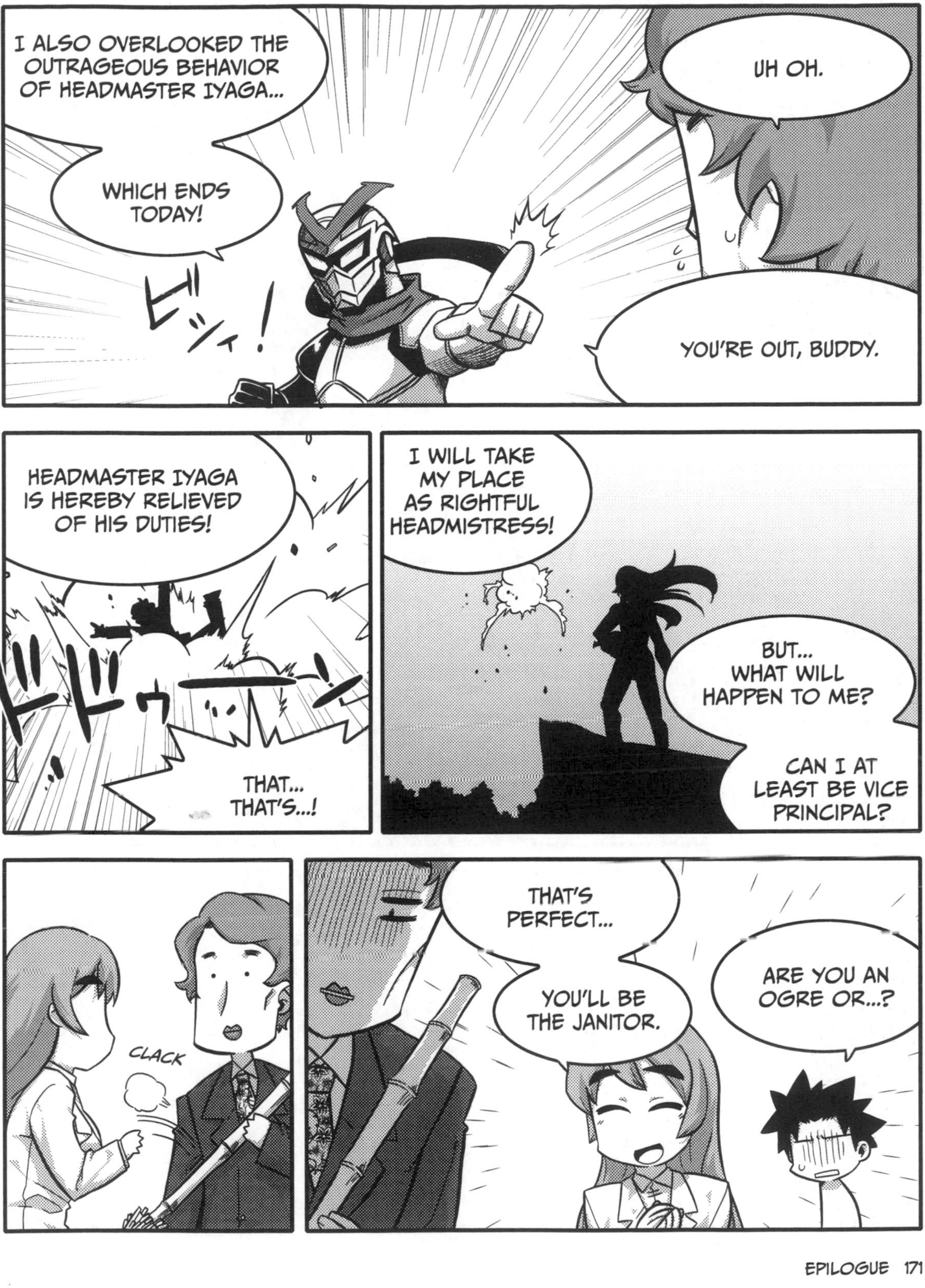
I ALSO OVERLOOKED THE OUTRAGEOUS BEHAVIOR OF HEADMASTER IYAGA...
WHICH ENDS TODAY!
UH OH.
YOU'RE OUT, BUDDY.
HEADMASTER IYAGA IS HEREBY RELIEVED OF HIS DUTIES!
THAT... THAT'S...!
I WILL TAKE MY PLACE AS RIGHTFUL HEADMISTRESS!
BUT... WHAT WILL HAPPEN TO ME?
CAN I AT LEAST BE VICE PRINCIPAL?
CLACK
THAT'S PERFECT...
YOU'LL BE THE JANITOR.
ARE YOU AN OGRE OR...?

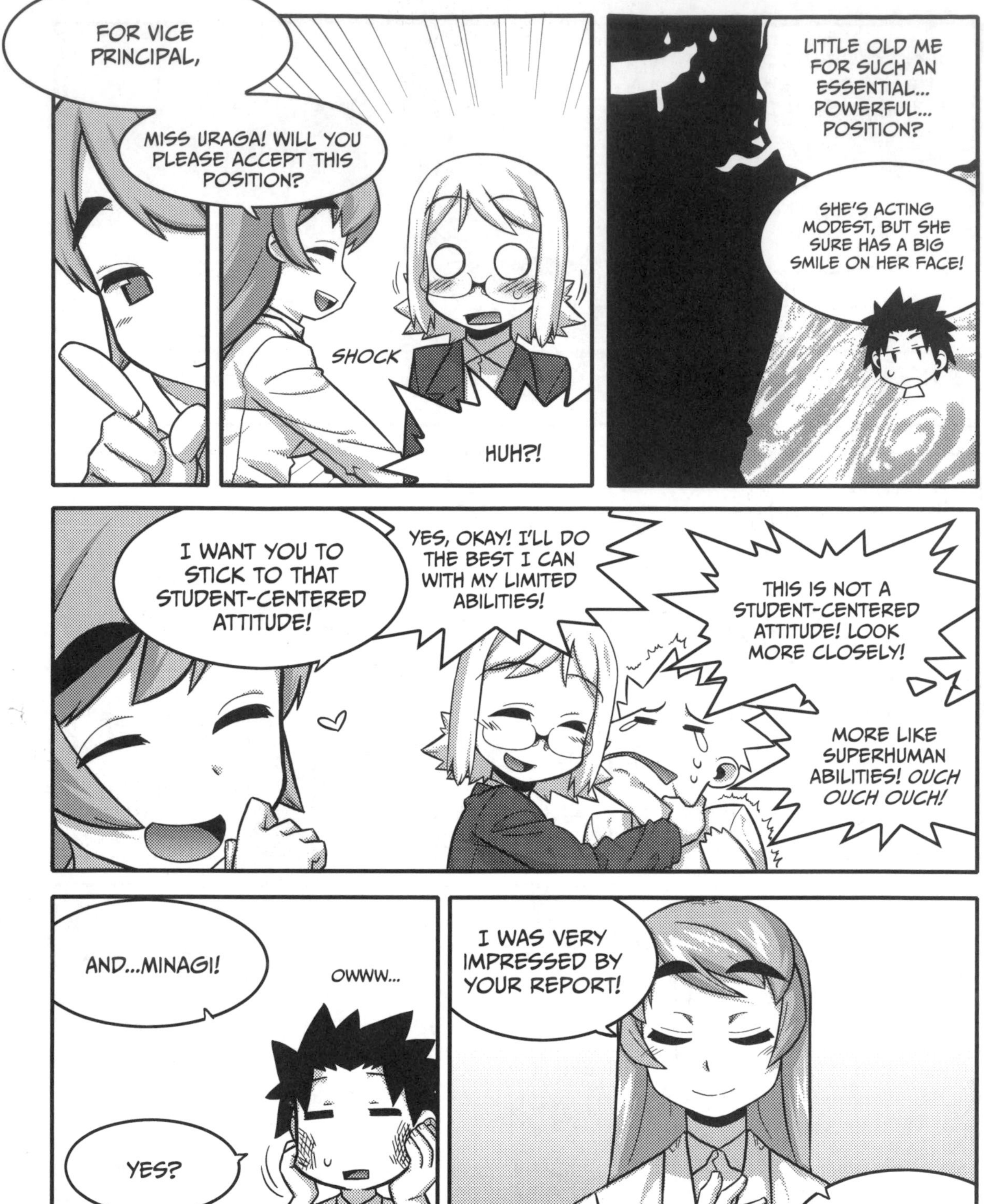
FOR VICE PRINCIPAL,
MISS URAGA! WILL YOU PLEASE ACCEPT THIS POSITION?
SHOCK
HUH?!
LITTLE OLD ME FOR SUCH AN ESSENTIAL... POWERFUL... POSITION?
SHE'S ACTING MODEST, BUT SHE SURE HAS A BIG SMILE ON HER FACE!
I WANT YOU TO STICK TO THAT STUDENT-CENTERED ATTITUDE!
YES, OKAY! I'LL DO THE BEST I CAN WITH MY LIMITED ABILITIES!
THIS IS NOT A STUDENT-CENTERED ATTITUDE! LOOK MORE CLOSELY!
MORE LIKE SUPERHUMAN ABILITIES! OUCH OUCH OUCH!
AND...MINAGI!
OWWW...
YES?
I WAS VERY IMPRESSED BY YOUR REPORT!
REALLY?

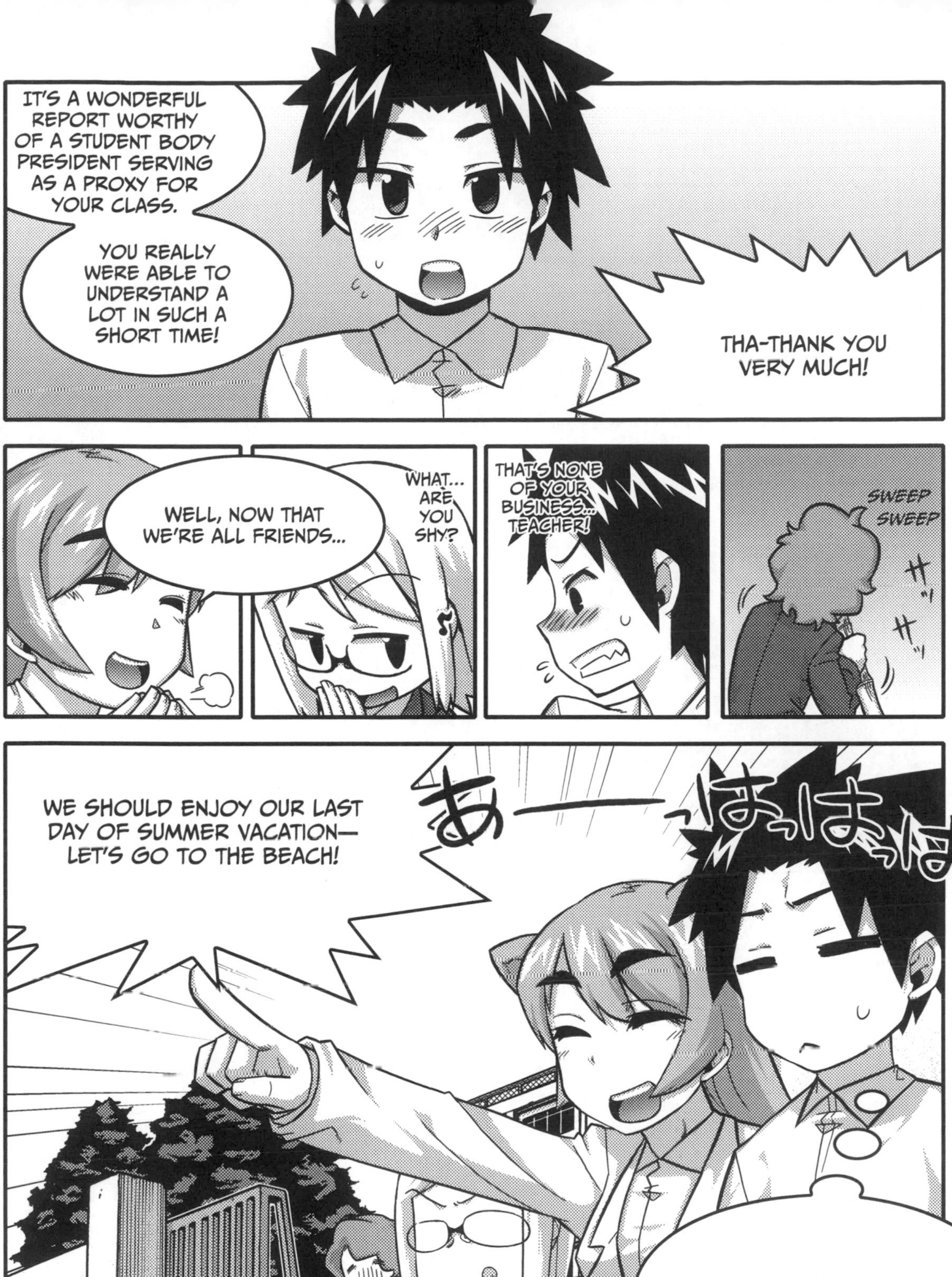
IT'S A WONDERFUL REPORT WORTHY OF A STUDENT BODY PRESIDENT SERVING AS A PROXY FOR YOUR CLASS.
YOU REALLY WERE ABLE TO UNDERSTAND A LOT IN SUCH A SHORT TIME!
THA-THANK YOU VERY MUCH!
WELL, NOW THAT WE'RE ALL FRIENDS...
WHAT... ARE YOU SHY?
THAT'S NONE OF YOUR BUSINESS... TEACHER!
SWEEP SWEEP
ザッ ザッ
WE SHOULD ENJOY OUR LAST DAY OF SUMMER VACATION—LET'S GO TO THE BEACH!
あーーっはっはっは
THIS SCHOOL IS SO WEIRD...

INDEX

AHEM!
TODAY, I WILL BE TAKING OVER AS HEADMISTRESS... HUH...IS THIS THING ON?
大概
THAT DOG CAN TALK?!

ABOUT THE SUPERVISING EDITOR

Hideo Nitta completed his doctorate at Waseda University Graduate School of Science and Engineering, 1987, majoring in Theoretical Physics and Physics Education. He is currently a professor in the Department of Education at Tokyo Gakugei University. He is also the author of *The Manga Guide to Physics* (Ohmsha, No Starch Press).

ABOUT THE AUTHOR

Masafumi Yamamoto completed his doctorate at Hokkaido University Graduate School of Engineering Division of Applied Physics, 1984. He is currently Representative Director of Yaaba, Ltd.

PRODUCTION TEAM FOR THE JAPANESE EDITION

Production: Trend Pro/Books Plus Co., Ltd.

> Founded in 1988, Trend Pro produces various media advertisements incorporating manga for a wide range of clients from government agencies to major corporations and associations. Books Plus is the service brand that takes Trend Pro's production know-how, which has achieved the best results in Japan, and specializes in book production. The brand is renowned for its preeminent professional team who provides complete design, editing, and production services.
>
> *http://www.books-plus.jp/*
> Ikeden Bldg., 3F, 2-12-5 Shinbashi, Minato-ku, Tokyo, Japan.
> Telephone: 03-3519-6769, Fax: 03-3519-6110

Scenario writer: re_akino

Illustration: Keita Takatsu

DTP: Mackey Soft Corporation